Voice Over Frame Relay

Voice Over Frame Relay

Learn

- ✔ **How to Save Up To 90% On Your Phone Bills And Still Get Toll Quality Voice**
- ✔ **How to Simplify The Administration of Your Network**
- ✔ **How to Carry Your Internet and Intranet Traffic As Well**

by William A. Flanagan

Published by Telecom Books
An imprint of Miller Freeman, Inc.
12 West 21st St., N.Y., N.Y., 10010
www.telecombooks.com

ISBN 1-57820-011-3

For individual orders, and for information on special discounts for quantity
orders, please contact:
Telecom Books
6600 Silacci Way
Gilroy, CA 95020
Tel:800-LIBRARY or 408-848-3854
Fax:408-848-5784
Email:telecom@rushorder.com

Distributed to the book trade in the U.S. and Canada by
Publishers Group West
1700 Fourth St., Berkeley, CA 94710

Transferred to Digital Printing 2007

August 1997

Dedicated to
all my family,
but especially those
who've been through
my book writing
and still put up with me.

Contents

Turning on a Paradigm: Fast Changes In Voice Transmission

Since Alexander Graham Bell invented the phone, the world has thought of voice transmission in the terms he created with the technology of his time. Bell's idea, the loop of wire passing an electrical current through two telephones, was so solid that for over 100 years we would think of voice transmission as a real-time process. One purpose of this book is to expand your view to include at least one other way to transmit voice: in data packets on a frame relay network.

Some History

Packetized voice first appeared in the 1970s. At that time the technology still mimicked Bell's current loop. The packets fully occupied a channel of fixed bandwidth assigned by time division multiplexing.

In 1991 FastComm Communications Corp. set a precedent when it was the first to demonstrat a voice connection over a public frame relay network. This was true packetization, with each voice frame taking its chances on a public data network. Other data traffic did cause unpredictable delay for the voice frames, particularly due to congestion. But the demonstration proved that variable delay, with its loss of a real-time relationship between speaker and listener, didn't destroy the sound.

This was new—and strange. Many people said then that it couldn't be done. They "knew" it was impossible to transmit voice over frame relay.

The members of the Technical Committee of the Frame Relay Forum (FRF-TC) originally shared that reluctance about voice. The FRF-TC is the group that quickly wrote and published the key implementation agreements (IAs) that make the booming frame relay market a reality. Voice Over Frame Relay, initially, had no such consensus. For several years I and a few others, notably Ted Hatala of the UK, kept bringing the subject up, and seeing it put off.

About 1992 a small group, often only 6 or 8 people, started working on a draft IA for voice. Charting new territory was slow; no one knew what VoFR should look like or what the scope of the IA should be. Then the first commercial VoFR products appeared. Their makers joined the working group. When the idea caught on (started to sell), the size of the working group doubled, then that tripled. The new people brought suggested text and new energy. And finally the VoFR IA, in the works longer than any other FRF document, neared completion early in 1997.

The draft sent in January 1997 for a vote by the full Frame Relay Forum carried a cover letter from the FRF Board, asking for the addition of another compression algorithm. This was an unusual procedure, but it was approved in the vote. The Technical Committee adjusted the draft at its March 1997 meeting and the FRF Board ratified it two months later. It is reprinted in Appendix C and its content is explained in Chapter 3.

In more than one sense, however, there is not all that much in the FRF VoFR IA—yet.

The only absolute requirement to claim compatibility with the first edition is to implement a small feature set associated one of two compression algorighms, in one of two classes:

1. ADPCM, according to G.727 (embedded ADPCM) with CAS.

2. the CS-ACELP voice compression algorithm defined in ITU recommendation G.729, with CAS signal transfer syntax (to convey on- and off-hook status) and DTMF transfer.

All other features are optional, to be used if acceptable to all connected parties. Since this IA applies only to permanent virtual circuits, which must be provisioned manually at installation, there is no great problem with configuring the VoFR devices at the same time. That's when you choose what kind of signaling to use, whether to support fax demodulation, and so on. If you do choose to implement certain features, then the IA says how you must do them in order to remain "compliant."

Physically it is a small document, barely a dozen pages without its annexes. To keep it small, the TC omitted most items that can be found in other standards documents. A lot of the details are assumed—like the order of bit transmission and which bit in a byte is the most significant (look in Q.922). I've included that kind of additional information so you can understand the VoFR process without looking in other documents.

We're Here To Help You

You'll want to understand VoFR as the next practical way to reduce your network costs. Savings from putting voice tie lines on frame relay data networks can pay back the additional investment in voice equipment in just months. Savings on international circuits are large enough to threaten present tariff structures (with some help from the FCC). From being "impossible," VoFR has become an essential element to be considered by network designers. This includes those who historically worked with data as well as those concerned primarily with voice.

From all the FRFTC meetings on VoFR, I've seen how VoFR brings together these two very different viewpoints. In pursuit of a common understanding, I've included material that one side will consider old-hat, but is probably not well understood by the other. My goal is to make it easy for anyone with a voice or data background to understand this new fusion technology, VoFR.

Reading This Book

For data people, this book includes information on voice interfaces and voice features. There are sections on traditional analog interfaces, digital voice formats, voice compression methods, and many kinds of signaling. Dyed in the wool voice people may know all this, and use these parts for reference only.

On the other hand, the reviews of frame relay and related data topics are done with the voice networker in mind. Data people may skim these sections.

Certain concepts have been included in both areas, to be sure everyone gets to see related text when dipping into the book for a specific reference. This book is not intended to be consumed in one sitting.

To help identify the jargon that may be new to you, certain words and phrases are set off in 'single quotation' marks. Those terms written with Initial Capitals should be indexed. "Double quotes" are used for their customary purposes.

Don't panic over unrecognized acronyms. There are hundreds used in communications, and many of them are defined in the glossary, Appendix A.

Note that an "Appendix" is part of this book. An "Annex" is part of the Implementation Agreement.

Where's ATM?

But wait: wasn't ATM supposed to be the fusion transmission technology for voice and data? How did frame relay take the job? ATM is obscured for most end users by the thousands of pages of specifications and a general "unfinished" feeling. In sharp contrast, all the FRF implementation agreements—the basis for today's frame relay market—fit in a 2-inch binder.

Relative simplicity is one of the reasons that attitudes toward voice over Frame Relay have changed even faster than we have come to expect in communications. Only 5 years after the first demonstration, voice over frame is routine. This book tells you what happened, how it happened, and how you can save serious money by combining at least some of your voice and facsimile circuits with your data on frame relay networks.

As always, your comments and suggestions are most welcome.

— William A. Flanagan
May 1997

Box 411
Oakton
VA 22124

Acknowledgements

My special thanks are extended to all the members of the Frame Relay Forum Technical Committee who have struggled with the Voice over Frame Relay Implementation Agreement. Each of you has contributed some insight, shared some fact, or helped me in some way to bring this book to reality.

With the large number of participants in the voice over frame relay working group, it is impossible to name everyone. However it seems essential to mention two people:

—Rajiv Kapoor, the first chairman of the Technical Committee, who helped keep the idea alive and under who's care the Voice over Frame Relay Implementation Agreement project got started; and

—Doug O'Leary, the second chairman of the TC and chairman of the voice working group, who saw the IA completed.

Thanks also to my coworkers at FastComm, who continue to encourage me in writing books, particularly our President, Peter Madsen. There would have been several errors and some significant facts missing if Jim Batteas hadn't given the manuscript a careful reading.

Thank you all.

— William A. Flanagan

Chapter 1

A Revolution In Voice Transmission

For over 100 years "the phone company" has provided voice service with very high expectations of service quality. First, phones work even when your local power fails. Second, the sound quality is consistently good. The technology prevents the fidelity from ever being "hi fi," but except in regions with difficult geography (and very long wires) the phone people always provide the best sound the system can deliver.

Technology is changing a lot of things, including these traditions among the successors to "the phone company."

Cellular service is a common example of changed expectations: it is not assumed to work all the time, and certainly not in all locations. The widely variable sound quality is tolerated even when it ranges from merely ok to outright awful. Requirements for sound quality and reliability that are rigidly applied to wireline phone service are simply ignored when considering cellular service.

Reduced expectations have opened the way to alternative transmission systems. Voice over the Internet may represent the most extreme example at this writing. Earlier attempts to send voice over X.25 networks were abandoned largely because the sound quality did not meet expectations at that time (and may never because of long, variable transmission delay across X.25 networks).

Voice over frame relay (VoFR) is a compromise. Compared to the best quality available from the Public Switched Telephone Network (PSTN), VoFR is usually lower quality, but not much lower. VoFR offers quality and reliability significantly better than voice on the Internet. We shall see that these issues of quality do not apply to "voice messaging" like voice mail, but only to near-real-time interactive calls.

Bell's Original Current Loop

The first telephone system deployed for a public service was based on a single continuous loop between two telephones (Fig. 1 is a simplified diagram). Essentially the same electrical current, from the battery in the central office (CO), runs through the transmitters and receivers at both ends of the connection. The local loops (wires) of the calling and called parties are connected to each other by the switch—originally a function of a manual cable patch board.

The first transmitter, a small container of carbon granules, is still in use today. When someone speaks into the phone, sound pressure waves from the speaker's voice fall on a diaphragm over the granules. A loud voice creates stronger waves that apply higher pressure on the diaphragm, which compacts the carbon particles. Pushing the particles together creates more paths for current (lowers the transmitter's electrical resistance) so more current flows through the loop.

Between pressure peaks in the sound there are pressure valleys that allow the diaphragm to bounce back away from the "carbon button." Looser particles have higher resistance which lowers the current in the loop.

At the receiver, an electromagnet converts the loop current into a force on another diaphragm in the earpiece. As the current fluctuates in response to the sound waves falling on the transmitter, the strength of the magnetic

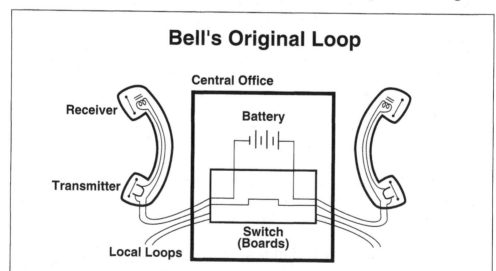

Bell's Original Loop

Central Office

Receiver

Battery

Transmitter

Switch
(Boards)

Local Loops

Fig. 1. Local Loops are extensions of the single loop that carried an electrical current between two telephones (simplified schematic). The phone company provided the current from a battery in the central office. First manual patch cords and later automatic switches created loops on demand.

field in the receiver changes in proportion. The springy diaphragm moves in response to the magnet's strength and thus reproduces sound waves like those that fell on the transmitter.

The speaker's own receiver sees these current variations too. Particular features of the circuitry in the phone keep the volume down, producing some audible feedback called sidetone.

Since there is only one electrical path to each phone (the two wires of the local loop), that path must also serve other functions:

— When you pick up a phone, go off hook, the hook switch closes and causes the phone to draw direct current (d.c.) from the CO. This tells the switch to supply dial tone and prepare to receive dialed digits.

— Before tone dialing, the rotary motion of the phone dial broke the circuit (stopped the current flow) as many times as the number dialed (0 causes 10 breaks). In England a rotary dialer is called, quite correctly, an "interrupter." It interrupts the flow of current in the local loop.

Inside the Telephone

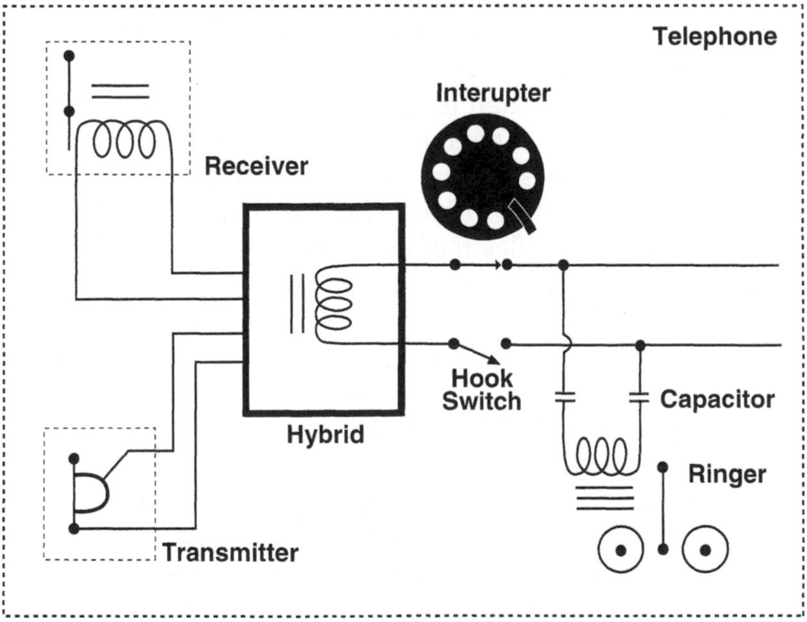

Fig. 2. In a telephone, the local loop connects separately to the ringer and the handset. Capacitors block d.c. put pass a.c. ringing voltage. The hook switch and pulse dialer control d.c. flow for signaling. The "hybrid" circuit, a complex transformer, couples the transmitter and receiver to the local loop.

— The switch announces a phone call by applying an alternating current (AC) to the loop. In North America this ringing voltage is 20 Hz and may be over 100 V in the CO. The phone ringer or bell inside the phone is attached to the loop through capacitors that block d.c. but pass a.c. current (Fig. 2). Ringing stops when you pick up the phone because the switch recognizes d.c. flow (through the hook switch in the answered phone) and shuts off the ringing generator.

— To end a call, you hang up, which opens the hook switch and stops the d.c. current. This tells the switch to clear the call.

From this review of how a phone works, it should be clear that the system has two distinct features:

1. A local loop can carry only one call at a time. Moreover, only one phase of a call is possible at one time. That is, you can't talk while dialing, and ringing occurs only while the phone is on-hook. In a sense, this is time division multiplexing, as each function takes its turn to use the loop.

2. Despite delay due to the finite speed at which electricity travels through copper, the two parties hear each other in "real time." That is, all events and speech at one end are reproduced in exactly the same sequence and with identical time relationships at the other end.

The first feature results from using a complete circuit (local loop) per call. Circuit switched connections like this consume all of certain resources which are dedicated to the connection. For example, the local loops at both ends of the connection are not available to other calls, nor are the connection points within the central office switch (if the call extends over several central offices, the same applies to the switches in each office and the interoffice trunks between COs). For that matter, the loop consumes electrical power, from the battery, that is not available for other purposes.

The second feature existed for so long that it became established as an iron-clad rule in the minds of many people. In particular, it is often the case that those who share the tradition of "the phone company," and those who have experienced voice transmission only on circuit-switched connections, place great importance on preserving the absolute time relationship between speaker and listener. This concern may be extended to include the timing between different parts of a speaker's transmission.

This expectation is not unreasonable. Most of the advances since the time of Bell have sustained this assumption. In the conversion of the public switched telephone network (PSTN) from analog to digital, the basic paradigm has preserved the main features of Bell's original architecture.

Channel Banks Emulate the Analog Loop

Everything that has been deployed in the digital version of the PSTN has been based on the assumption (implicit, seldom explained) that all the features of Bells' original "single loop between phones" must be preserved. The justification is that new equipment must interoperate with earlier phones still in use.

What this means is that the PSTN, in all its advances to date, has accepted the necessity to dedicate portions of the network to each call. These dedicated portions are no longer available for other calls. One of the main reasons that this assumption has not been challenged is the implicit acceptance of feature number 2, the fixed relationship in time between the sender and receiver. Preserving time relationships has required dedicated channels that spend at least half their capacity on "filler" or idle signals to preserve relative time.

Less than half of the bandwidth is actually used transmitting voice:
— a speaker normally is silent while listening;
— both parties may be silent at times;
— averaged over all callers, each end speaks less than half the time.

At the very start of the evolution from analog to digital in the public network, an example of this preservation of features was seen in the channel bank. The CB is the first example of digital voice transmission equipment (Fig. 3). It was introduced in the 1960s to save copper loops between central offices: two wire pairs carry 24 conversations, a gain of at least 12:1 for simple phones, and a much higher pair gain for more complex interfaces, like E&M used on voice switches and PBXs. E&M can require as many as 8 wires, leading to a reduction of 48:1 in wire count.

The digital portion of the transmission, between channel banks, emulates the activity on the analog local loops on either side. Each loop current,

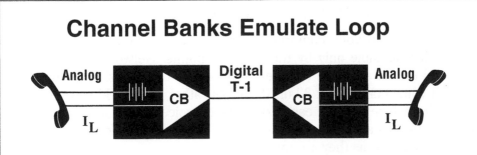

Fig. 3. Channel banks introduced digital transmission to the public telephone network in the 1960s. A voice circuit between channel banks preserves the time relationship between speaker and listener by encoding the loop current in a flow of digital bits over a dedicated time slot on the T-1.

IL, is passed in digital form (1s and 0s) over the T-1 line in the center (between channel banks) to the other analog loop. There is no impact on an analog phone or other telephone equipment, which continues to see the standard analog loop current as if there were a single physical current loop between the two ends.

Note that the bit flow between the channel banks replaces, but does not change the nature of, the analog functions at each end. A time slot on the T-1 (defined as $1/24$ of the bandwidth on the T-1 line) replaces a pair of copper wires in the original loop. However, the digital implementation (a bit stream to encode voice amplitude plus signaling) performs the same functions as the analog current.

Digital Hierarchy Preserves Emulations

When more than a few T-1 lines (also known as digital signal level 1 or DS-1) are needed between two central offices, the modern practice is to multiplex them onto a line at a higher bit rate. Most of the transmission equipment deployed until the last few years runs at 45 Mbit/s, a digital signal level 3 (DS-3) circuit. A DS-3 carries 28 T-1 circuits that are combined in an M13 multiplexer (1 and 3 indicate multiplexing between levels 1 and 3).

Lately, optical fiber has replaced copper between COs. Fiber optic multiplexers often start at 52 Mbit/s on the "slow" side and combine many such inputs onto an optical signal at 155 or 622 Mbit/s. The format on the fiber does not matter—it could be Synchronous Optical Network (SONET) or proprietary. When a DS-3 goes in at one end, it comes out a DS-3 at the other.

In other words, the SONET or optical transmission line simply preserves the underlying T-1 signal end-to-end, and again preserves the emulation of the current loop.

Compressed Voice in TDM Channels

Early forms of voice compression produced a continuous bit stream that was understandable only by the two end points. Network designs simply assigned narrower digital channels (defined by time division multiplexing) to each voice connection: 32 or 16 kbit/s rather than 64 kbit/s were common. The nature of these early compressed voice signals are ideally suited to T-1 networks and digital leased lines. They helped make private T-1 networks as popular as they are today.

Around 1980, compressed voice information was first put into packets that might be recognizable as a standard format (HDLC, for example). The packets provided error checks, a place for signaling information, and some management functions. However, the bit stream within the packets was still continuous as far as the encoder and decoder were concerned.

So, either packetized or as an undifferentiated bit stream, compressed voice in TDM channels is another emulation of the copper loop. Compression simply reduces the bandwidth or the number of bits needed.

ATM Specification Is More Loop Emulation

The ATM Forum has defined what it calls "voice over ATM" but is, in basic terms, an emulation of a T-1 line across a "cell relay" network (Fig. 4).

In this case, the ATM network transports the T-1 bit streams in an emulation of wire lines or TDM multiplexer channels. That T-1 still emulates an analog loop whether it is carried on copper pairs, TDM multiplexers, or a cell-based transmission system (like ATM).

The practical implication of T-1 circuit emulation is that the ATM network carries the same number of cells per second when all 24 voice channels on the T-1 are idle as when they are all busy. No benefit is derived from the statistical nature of voice traffic or the statistical multiplexing of cells from different voice connections.

Inefficient use of the ATM network will be acceptable for some time (and even necessary for phone companies who need to provide T-1 service over their ATM backbones). But eventually the extra cost should drive telephone companies, particularly long distance carriers, to eliminate cell transmission when a voice channel is idle or silent. At this writing, the first

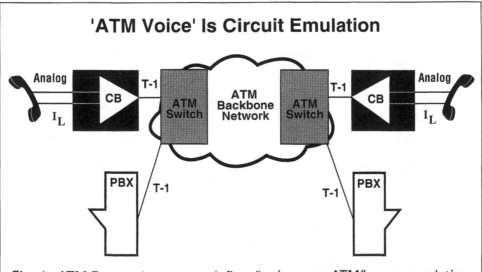

Fig. 4. ATM Forum documents define "voice over ATM" as an emulation of the T-1 circuit (circuit emulation) which emulates the analog loop.

work for dealing with packetized voice on ATM has been started by the wireless (cellular) working group. These companies pay real money to other carriers for lines to connect cell sites. An ATM service makes sense for this application only if it can be more efficient than circuit emulation.

Under circuit emulation, the ATM network knows nothing about the type of traffic on the T-1s being carried. Traffic could be broadcast video or data for stock ticker reports just as easily as voice connections. When an ATM net carries a T-1 without regard for its payload, the net must transport all 1.544 million bits per second even if the bits mean nothing.

A New Paradigm in Packets

But "what if" the network did know about the traffic content? Or, what if at least the access devices could tell if a voice circuit were idle?

Then the backbone network would be relieved of the burden of transmitting frames or cells (or any form of bits) when no one was speaking.

Data Does It

This problem has been solved by various kinds of packet switches for data—like Local Area Networks (LANs), statistical multiplexers, front end processors, and others forms of data packet handlers.

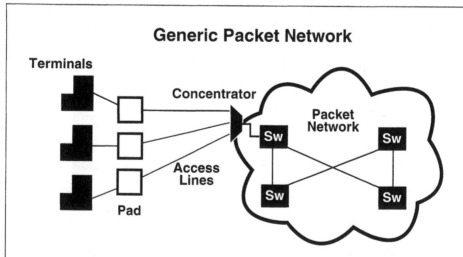

Generic Packet Network

Fig. 5. A generic packet network illustrates the key features: users may be permanently attached on physical media; null characters fill time between data packets; concentrators and switches discard received nulls; all users share switching and transmission capacity, taking turns to put packets on aggregate lines.

In a generic packet network (Fig. 5) the switches, trunks, access lines, and concentrators are shared by user's terminals. When a user device has data to send, it formats a packet and transmits it on the access link to the concentrator. That's simple to say, but it implies that the user device understands what is "real" data and what is not, so that only the "real" information is included in a packet. More than that, the user device needs to know where the data should go, so the proper address may be inserted into the header. The header is control information added as the first few bytes of the packet (Fig. 6).

Inside computers, the software keeps track of what's data and what's not. By contrast, a normal telephone is too simple (dumb) to tell the difference between the latest news and the whistling of a tea kettle—both are sent without distinction.

Between packets, when there are no data, the packet access device sends null characters to preserve link synchronization. Nulls are discarded by the device that receives them at the other end of the link. Nulls are not switched through the network.

The concentrator can receive packets from all attached user devices on all access links simultaneously. Data are stored temporarily in memory, so the total input rate may exceed the aggregate line speed, at least for a time that depends on the amount of memory in the concentrator.

When the aggregate line into the network is available, the concentrator selects a packet and transmits it. When that packet is finished, the concentrator sends the next packet in queue. If no packets are queued to send, the concentrator fills the aggregate line to the switch with nulls. Nulls never occupy bandwidth unless there are no data waiting.

Note that the aggregate line speed may be much higher or much lower than an individual access link speed. For example, a stat mux with many 9600 bit/s inputs and a 56 kbit/s aggregate is common. But the opposite is

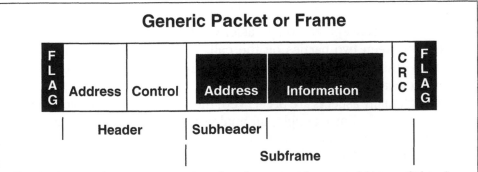

Fig. 6. Frame format shows header that contains an address of the destination. The information payload is only "real data" as identified by the sending device. The control field may serve many functions.

also true, as when a router has many LAN ports for access (perhaps more than 50 Mbit/s total) and only one 1.5 Mbit/s aggregate or less going to the wide area. Yes, this does mean that a LAN router is a form of packet concentrator or even a packet switch.

A switch directs packets among many aggregate "trunks." The proper route is determined from the address in the packet header and a list of destinations (the routing table) held in each switch. When a user is not sending data packets, the only resource consumed for a connection is memory space for the related entries in the routing table.

Recall that a telephone-type circuit switch devotes considerable resources (bandwidth in a time slot, local loop, electrical current, etc.) exclusively to each voice connection, even when both speakers are silent. That is, needed resources are the same to prepare the network as to actually carry information.

Packet data systems, by contrast, don't consume bandwidth unless there is a real need. That's because the end devices are smart enough know what constitutes data. Voice over Frame Relay (VoFR) applies processing power (intelligence) and many of the same packet principles to get similar benefits for voice.

The "catch" or cost comes from the need to process analog voice signals far more than a channel bank does or a telephone can. Serious computing power (in the form of a Digital Signal Processor, DSP) is needed to compress the bandwidth requirement, to know when a channel is idle or silent (to be able to suppress transmissions), and to cancel echoes in a digital form.

The benefits to users are drastically reduced costs for voice circuits and the simplification of the networking job by combining voice and data on one backbone.

Beyond Bell's Voice Concepts

Bell's loop concept was very basic, very physical, based on "real" connections between two parties on the call. The ability to respond to traffic content is outside of or beyond an architecture based on a loop of electrical current dedicated to each conversation.

Though many aspects are preserved, the new paradigm for voice is quite different (Fig. 7). For example, an analog voice call almost always is converted first to exactly the same digital form a channel bank uses: pulse code modulation at 64 kbit/s.

Compression is the next essential element. The DSP may apply any of a dozen standard algorithms or compression processes. There are many more proprietary methods. The result is a bit stream that carries a description of the speaker's voice, but at a lower bit rate. Standard compression algorithms produce outputs of 32 kbit/s down to around 8 kbit/s. Proprietary formats can reach below 3 kbit/s.

Comparing these encoding rates to 64,000 gives what is commonly called the compression ratio; voice encoding at 8 kbit/s is 8:1 compression, for example. The effect of silence suppression usually is not counted in this ratio.

Voice compression consists of making approximations and omitting some information, perhaps about higher frequencies. Many algorithms rely on predictions, based on normal speech patterns, calculated at both ends of the circuit. Sending only the variation from the prediction, not the full description, saves bandwidth but requires more processing at both ends.

At this point the reduced rate voice signal could be sent over a narrow channel. In fact, most compressed voice found in private networks is exactly that. Just as a normal voice conversation consumes dedicated resources, so does compressed voice when sent as a stream that fills a channel. At its base, this is another form of loop emulation.

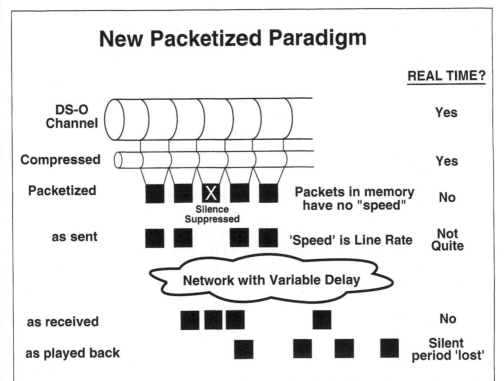

Fig. 7. Packet voice processing starts with a familiar step, conversion from analog to digital as PCM at 64 kbit/s (A). Compression reduces the bit rate (B). Packetization (C) prepares the information for the backbone network.

Packets Are the Key

The next important advance is to divide the compressed bit stream into relatively small groups of bytes, often called frames. In the context of voice over frame relay, the term "frame" will be preferred here.

This division may appear simple, but some compression algorithms need to have bits delivered in a predictable fashion. For example, PCM needs a dedicated channel. Variability in delay across the network causes problems for systems designed to fit dedicated channels.

Often a problem lies in the way the receiver calculates the predicted loudness level; delayed information throws off the prediction and the process may be slow to recover. Therefore the algorithm must be optimized for frames in order to tolerate delay ('latency') and variation in delay ('jitter').

Once the bit stream describing a voice sound is cut into segments and assigned to frames, there are more options on how to handle it. For example, the DSP that does the compression knows if the sound level is low throughout the time period compressed into a given frame. If so, that period is considered "silence" and the frame might be discarded.

Smart access equipment treats voice frames as high priority. They are sent before data packets in order to minimize the variation in delay for voice information as it travels from speaker to listener. However, a complete voice frame sitting in a memory buffer is almost like a short voicemail message. It can be retrieved and listened to at the user's convenience. In live conversations, the network and access equipment try to make that as soon as possible, but it does not have to be exactly synchronized, as we shall see in Chapter 3.

But What About...?

From the mid-1980s, when frame relay was introduced, network operators hesitated to use frame relay at all, even for data. Frame Relay was new, unproven—but mostly it was not understood. Until 1995 there was even more reluctance about voice over frame relay, all the way to "It can't be done!"

Let's examine some potential objections and lay them to rest. Chapter 3 describes the details.

...Data Loss

Note that the nature of frame relay (as defined in the standards) is a best-efforts service. Carriers commit to deliver an average data rate. There can be no guarantee for a particular frame. Any frame may be discarded if a switch is temporarily overloaded or there is an equipment failure. An error in a frame causes the receiver to discard it.

This potential to "lose data" has scared away some users. Also, carriers don't want to be accused of throwing away a customer's data. But these concerns are overblown and ignore history.

In reality, any network—packet or circuit switched, analog or digital—will introduce errors.

Any transmission medium is subject to errors. In the 1980s, analog lines were considered acceptable if the bit error rate was less than 1 in 1000. Such lines would have been nearly useless for asynchronous data terminals (which cannot themselves correct errors) if modems did not include software to correct line errors.

Synchronous digital lines don't have error correction in the CSU/DSU or the network. That is because all modern synchronous protocols include a functional layer that corrects transmission errors. The Transmission Control Protocol (TCP) protects against errors that occur in the Internet Protocol (IP). IP is a "best-effort" or datagram service with no more delivery guarantees than Frame Relay. Certainly no one hesitates to use the Internet because of the relatively high potential to lose a given data frame. TCP makes it work.

Data error correction, like TCP, relies on the sender saving a copy of each data packet until the receiver acknowledges accurate receipt. If not, the sender can repeat the transmission as many times as necessary. So data are never "lost" or "thrown away" as long as the sending protocol processor keeps a copy until it gets through. With the wide adoption of error correcting protocols, frame relay for data is as reliable a transmission medium as any other.

Correcting errors by retransmission takes time, too much time for voice. VoFR may apply other strategies to deal with lost frames. But mostly this new voice technology relies on the vastly improved reliability of frame relay networks and their ability to deliver the original transmission.

The optical cable and new electronics in frame relay switches have lowered the expected bit error rate before correction to better than 1 bit in 100,000,000,000. An extended error test over a large geographic loop on a public network a few years ago showed no bit errors at all over a weekend, about 60 hours.

...Maintaining the Timing

Once packetized and sent from the access device to a concentrator, the speaker's voice information has lost its tight timing relationship with the speaker. Jitter tends to build up as the number of switches in the path increases. As the next two chapters will explain, there are ways to deal with jitter, loss of frames, and other attributes of a frame network.

To help understand the source of this concern, consider a telephone professional who has had an entire career based on Bell's technology—the "real time" current loop architecture. Then introduce a new voice transmission method based on frames sent over a network that makes only a "best effort" to deliver information and has no real control over the exact delivery time of each frame. The differences have been known to induce culture shock.

The first reaction was flat disbelief: it is impossible to transmit voice over frame relay. Years of demonstrations at trade shows finally turned that around.

Then it was "OK, you can, but not with acceptable quality on a busy network." Clever use of committed capacity, giving it to voice with a higher priority than data, answered that concern.

Another natural reaction of someone from the Bell background was to attempt to emulate a loop again. Many features were proposed for different parts of a frame relay network to control the generation, transmission, and playback of voice frames in an attempt to keep them closely synchronized:

— Time stamp all voice frames when created so the receiver knows when to play back each one.

— Modify all the frame relay switches in all the public and private networks in the world so they can treat voice frames (packets) with special priority in order to reduce delay and jitter.

— Receivers will play out voice frames according to the time stamps, and discard any frame that arrives too late to be used at the "proper" time.

There are problems with implementation here. Start with the frame relay standards: they define only one type of frame. The traffic management features to control class of service (committed information rate, DE bit, burst sizes, etc.) apply to all frames equally.

The cost to refit all frame switches with new features to distinguish among new categories of frames would be huge. Worse, it would take years to define, in standards and implementation agreements, exactly how to modify the switches to preserve interoperability among everyone's equipment and services.

After a few years of debating, everyone on the Frame Relay Forum conceded that the networks won't change much, and the standards will remain as they are. However, individual switch vendors may implement proprietary ways to assign priorities in the backbone. Makers of access devices have been free to optimize in proprietary ways within the range of implementation agreements and standards—and they have, as we shall examine later.

In short, frame relay will have to be taken as it is. Voice will have to be dealt with almost entirely in the access device.

...Waiting For Data Frames

Data packets and voice frames get in queue at every transmission point before they are sent on an outbound link. The length of that queue ahead of a voice frame, and the bit rate of the line, determine how long the voice frame has to wait. The number of bits waiting, divided by the line speed in bits/second, is seconds to wait. (That's because the "bits" cancel when you divide. See, you still need the fractions you studied in 7th grade.)

The scare here is that a voice frame is going to be queued behind a very long data frame, waiting to use a slow link. Some Token Ring frames may be 30,000 bytes long, which is a lot longer than the 20 to 50 bytes of a voice frame, and so will take a lot longer to be sent.

All very true, relatively speaking. But putting some numbers to the concept makes it less scary. Some typical and extreme values are collected in Fig. 8.

In addition, note several other assumptions:

— Time interval represented by a voice frame: 20-30 ms.

— Time delay to compress 20 to 30 ms of voice information into a frame: 5-40 ms.

— Link speed between frame relay switches: 1.5 Mbit/s to 45 Mbit/s.

— Local access link speed: 56 kbit/s to 1.5 Mbit/s.

If the frame relay backbone has 45 Mbit/s trunks between switches, the delay of waiting for the largest possible data packet is only 5 ms. Even better is

Packet Delay Calculations

Link Speed, kbit/s	Packet Size, Byte	Duration, ms
45,000 (DS-3)	40	.009
	256	.05
	30,000	5
1,500 (T-1)	40	2.6
	256	1.4
	30,000	156
56 (DS-0)	40	7
	256	37
	30,000	4300

Fig. 8. Typical values and calculations for packet accumulation delay. This is the time to transfer a packet or frame of this size across an interface at this speed.

the prospect that the really high-speed backbones will all migrate to ATM, or cell relay. In an ATM network, all frames are cut into pieces that fit in 48-byte payloads of 53-byte cells so there will be no possibility of a long data frame delaying a voice frame. This means that unless the backbone is not ATM, has slow trunks, or the long frames are going to the same low-speed access loop as the voice frame, queuing delay variation should not be a problem.

That last condition is the most valid concern: long data frames mixed with voice on a 56 kbit/s access. That delay is unacceptable by any measure except voice mail. The solution is in a draft FRF implementation agreement (IA), FRF.12, to be required by the VoFR IA. The draft IA on fragmentation defines how an access device divides and reassembles long data frames when voice frames are present at either end of the connection.

Fragmentation was separated from the VoFR document because data-only FRADs (frame relay access devices) will have to fragment and reassemble data frames if any of their virtual circuits or connections end up on a VoFR device. There would be little gained if other locations were allowed to send unfragmented data to a 56 kbit/s access point that also handled voice. But the delays from waiting for a 256 or 128 byte data fragment are manageable—much less than the time delay of a satellite hop.

Cost Analysis

Packetized voice over frame relay confronts the largest cost component of corporate networks: the monthly charge for lines. Carrier charges may represent 80% or more of the lifetime cost of a network (from installation to replacement). By comparison on that basis, equipment and labor are relatively small costs.

Private lines traditionally carried a monthly charge based on their length, speed, and location. X.25 networks charged for usage by the byte or packet. Internationally, the rates can be staggering to managers accustomed to U.S. prices.

Frame relay has been different. Perhaps due to a lack of ready ability in the switches to record usage, frame relay service is offered mostly at a flat rate per month. The base charge, related to the speed of the access link, has been lower than a leased line of that speed. This advantage starts at relatively short distances, though it is greater over longer distances where the leased line mileage is higher.

Some carriers also offer a variable grade of service quality based on a Committed Information Rate (CIR), and burst capacities (committed burst, Bc, and excess burst, Be) associated with a virtual circuit. The sum of CIRs on a physical link is often restricted to no more than the link speed. If the user stays at or below the CIR, the carrier agrees to deliver all frames and not discard any under normal conditions, even when the network is congested.

Burst sizes determine how long a time the sender can fill the access link with bits. Remember that frames can be sent only at the line speed, not the CIR, because all of a frame must be presented to the line without interruption. Then the next frame can be transmitted, or the sender may pause by inserting extra idle characters (HDLC flags) between frames. Each frame, then, is a burst, a certain number of bytes or bits. Several short frames may fit within the burst limit, commonly measured over a time interval of 1 second. Beyond the burst limit, frames may be marked as low priority (discard eligible, where the DE bit in the header is set) or simply discarded immediately.

Local tariffs vary widely, so a calculation should use the real costs for local access at each site plus long distance charges. An article by Steve Taylor in *Data Communications* magazine did that for an example of a long distance network only, assuming frame relay access at each end was no more expensive than regular phone lines. Costs of voice hardware were based on averages. This choice is potentially misleading, but probably conservative going forward as equipment prices are dropping.

For the example, voice calls on a virtual private network were priced for a small to medium size company at about $.09 per minute. These calls on the public network would have all the quality and reliability characteristics we expect from "the phone company."

Frame relay voice circuits were priced as additions to an existing data network. That is, the access line was in place, but the voice equipment, bandwidth, and virtual circuits were added.

In the example, conversations (or fax transmissions) over frame relay virtual circuits cost about $.0025 per minute. If the example had included international circuits, the charges would have been higher all around, but savings per minute would have been greater too.

The bottom line: in the right situation, frame relay can offer very inexpensive voice connections. Since saving money is the main driver for companies to install new technology, voice over frame relay will get considerable attention into the 21st century.

Chapter 2

Voice Digitization And Traditional Compression

The first digital treatment of voice, the channel bank described in Chapter 1, emulated the copper wire loop. Even when voice was compressed after digital conversion, it remained for many years a continuous flow of bits—like the current in the loop.

This chapter will describe the digital technologies used in circuit-switched connections in the widely deployed time division multiplexer (TDM) networks. The next chapter will cover packet-switched voice, that is, voice carried as data packets or frames on logical connections rather than in TDM channels.

In the last five years there has been considerable development in the understanding of voice and how it should be digitized. The first methods focused only on the sound source. Encoding attempted to capture in digital form what was happening to something physical and objective, like sound pressure or current in a wire loop. Encoding started with simple, direct measurements (see Pulse Code Modulation). Next came features that took advantage of statistical properties in voice, for example the predictor function in ADPCM, to represent the same type of measurements with fewer bits.

The latest techniques certainly record the physical reality. But how the encoding is done may reflect how we hear as much as how we speak. The listening process affects what a listener understands from a sound, and certain encoding methods are based on characteristics of the ear. For example, many encoding schemes use variable sound filters, like multi-band tone controls. They help encode sounds and shape reproductions by corresponding to the frequency response of the ear.

A third way to describe sound is to create a model of how the sound is produced: frequency, loudness, resonance, etc. Then the encoding of a sound is just a list of the parameters or variables in the model. A model may be quite complex, and subtle, but have only a few parameters. Those values may be transmitted in far fewer bits than the straight-forward PCM representation of the same sound.

Before moving forever into the digital realm, it is worth while to recognize the accomplishments of analog transmission designers. For over 100 years the public telephone network was analog—not digital. Yet there were huge multiplexers, high speed transmission lines, and clever procedures to save bandwidth on long distance voice connections. All of those features are easily associated with the digital era, yet they also existed before the transister, when all voice transmission was still analog.

Time Assigned Speech Interpolation (TASI) was used on transoceanic analog cables to take advantage of the one-way (half-duplex) nature of most phone conversations. Smart switches at each end of the cable temporarily assigned an analog channel to a conversation only when the speaker's voice volume exceeded a threshold. Early switches, being slow by digital standards, needed large groups of trunks, 100 or so, to make the statistics viable. A successor technology, digital speech interpolation (DSI), was deployed before packetized voice was introduced.

Neither TASI nor DSI worked at all with data modems or facsimile. TASI and DSI couldn't cope because the statistics changed—fax never stops, data never gives up the channel by going silent.

TASI, DSI, and similar methods relied on the stop-start statistics of human speech. Digital voice compression uses these properties also, but takes further advantage of many additional, more subtle features in voice.

Recently, conventional bandwidth saving has focused on using fewer bits (compression) in a slower or narrower channel (Fig. 9) rather than attempting to juggle time slots or TDM channel assignments. Techniques have improved steadily, but are still based on an initial conversion to pulse code modulation (PCM), introduced in the 1960s. PCM is the quality standard and often the first step before additional processing.

Pulse Code Modulation

The existing worldwide standard for digital voice is called Pulse Code Modulation (PCM). Under good conditions, the quality of this method of voice transmission can be excellent. It is praised as 'Toll Quality' because it first appeared on interoffice and toll trunks between telephone company central offices. These trunks have the strictest quality requirements.

Digitization

The PCM conversion between analog and digital can be done in one step, within a single integrated circuit chip, the codec (COder-DECoder). Traditionally, it is done in two steps (Fig. 10).

1. **Pulse amplitude modulation.** The incoming analog signal, representing the variations in a voice, is sampled 8,000 times per second. The modulator uses the sample to send a very narrow square wave pulse whose voltage (height, or amplitude) is the same as the analog signal's at that point. Imagine a board fence with a curved top. Take out most of the boards (leaving only every tenth, say). The original curve of the fence top can still be seen in the "samples" represented by the remaining boards.

2. **Digital encoding**. The height of the pulse is then converted to a digital value by a coder, an analog to digital converter. The output is an 8- to 20-bit code word (hence "pulse code") representing the voltage of the pulse and thus the analog input at the time of the sample.

The one-step PCM process converts an analog voice signal to a digital stream of 64,000 bits per second (bit/s); 8 bits per sample x 8,000 samples/sec.

The rate of 8,000 samples per second comes from the Nyquist theorem. This theorem shows that an analog reconstruction from digital data can contain all the information of the original analog signal—if the sampling rate is faster than two times the highest frequency in the original signal. In other words, if enough fence boards remain. Technically, sampling must detect

TDM Voice Compression

Fig. 9. Compressing voice to fewer bits per second allows the connection to fit in less bandwidth, but the constant bit flow at a fixed rate again emulates a current loop.

every change in direction (up to down or the reverse), or every change in sign of the analog signal (positive to negative or the reverse).

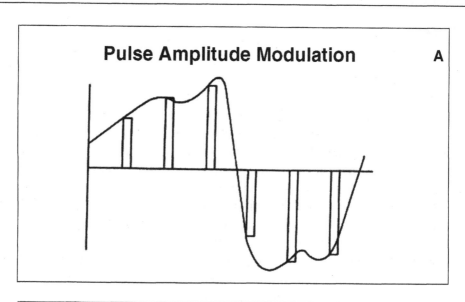

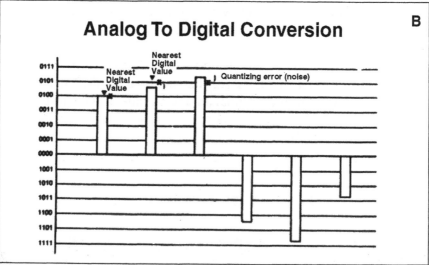

Fig. 10. Samples of the analog input are taken at 1/8,000 second intervals. Sample values are converted to pulses (A), a process called Pulse Amplitude Modulation (PAM). Heights of pulses are measured in digital numbers (Analog to Digital conversion). The digital value is what is transmitted as a stream of binary bits. Received bits are converted back to numbers, then pulses. Pulses are filtered to recreate the analog wave.

If sampling is not rapid enough, the resulting digitized points can represent more than one analog signal. This phenomenon is aliasing (Fig. 11) and produces unintelligible sounds.

To avoid aliasing, voice inputs are low-pass filtered to block any appreciable amount of signal at a frequency above 4,000 Hz. The filter adversely affects adjacent frequencies. The usable upper limit is only 3,300 Hz. Filtering out the low end, to block 60 Hz hum from power lines, puts the practical lower limit at 300 Hz.

The size of the sample, 8 bits, was determined after considerable experimentation, and a large amount of invention. The problem was to optimize the trade-off between bit rate and voice quality. It didn't hurt that computers then were starting to deal with 8-bit characters.

An analog signal, by definition, has infinite variability—it can take on any value. The digital representation of the same signal can take on only a relatively small number of discrete values, limited by the number of bits per sample: 8 bits allow 255 values. Therefore, at the precise time of a sampling, the analog input is seldom exactly the same as one of the possible digital output steps. The CODEC, however, must make a selection, and will pick the closest digital value. The difference between analog input and digital measurement (between the dot and the X in the PCM drawing above) is digitizing distortion, or quantizing noise. The human ear is very sensitive to quantizing noise. The distortion sounds bad. The quantizing process can be compared to someone measuring the height of the boards in a fence to the nearest foot when the length could vary an inch.

Early listening tests showed that if the analog input were measured with many digital output values very close together, the quantizing noise

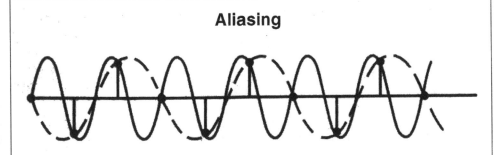

Aliasing

Fig. 11. Too few samples of a wave allow the resulting digital points (dots) to represent more than one signal. Aliasing is avoided by filtering input signals to eliminate high frequencies. Filtered voice signals must have nothing higher than half the sampling rate of 8,000 times per second, or 4,000 Hz

could be reduced to where it was not important. Unfortunately, the number of digital values required to cover the full volume range of a voice signal in such small steps is at least +/- 2,000. This is like measuring the fence height in millimeters. To number that many steps requires 12 to 16 or more bits per sample. At 8,000 PCM voice samples per second, that would be at least 96,000 bit/s.

"Hi Fi" codecs in stereo equipment may use 16, 18, or more bits per sample, and 44,000 samples per second. That's per stereo channel. CDs sound better than a telephone. The price is higher bandwidth: not very hard to get if you stay in one box; too expensive for telephone calls.

Even 25 years ago, when T-1 was introduced, designers recognized the possibility of compressing voice. They simply gave less attention to the very loudest levels. That is, by concentrating the digital measurement steps in the low and normal volume range, they reduced the number of steps needed for "toll quality" to 256 (an 8-bit binary word). In effect, the quantizing noise was kept very small at low volume levels, but allowed to increase with loudness. The effect is masked by other distortions created by the microphone, receiver, and lines when the volume is high.

For simplicity, the first analog to digital conversion is linear, into a binary number with 12, 14, 16, or more bits. Then the processor converts that large binary number to an 8-bit number by using a conversion table.

To concentrate the measuring at the low end of a range produces a highly non-linear ruler (Fig. 12). To measure a fence with it, some graduations might be 1 mm apart; others, as much as 1 foot apart. Voice engineers designed a non- linear "voice ruler" with the "fineness" of an adequate linear rule (16-bits)

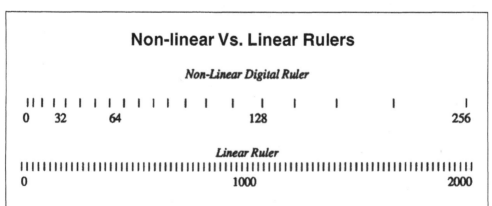

Non-linear Vs. Linear Rulers

Non-Linear Digital Ruler

0   32   64   128   256

Linear Ruler

0   1000   2000

Fig. 12. To concentrate on the low end means to be more exact with low signals than with loud passages. To do this the pulse height is measured in a non-linear way. For example, it can be compared to measuring a fence with a non-linear ruler.

near zero, and wider spacing at louder levels. This technique needs only 8 bits to measure pulse heights over the full volume range of a voice.

In other words, a non-linear voice encoder saves at least 33% of the digital bandwidth. The original signal is compressed for transmission, then expanded at the receiving end to the full 14- or 16-bit range. The two-phase process is known as companding. Thus PCM, today's standard, is itself a form of voice compression.

The specific form of non-linearity is the 'mu-law' algorithm in North America and Japan (T-1 regions), 'A-law' in the rest of the world (where E-1 is used). Differences arise in how the linear ruler is segmented to correspond to the nonlinear ruler (which ranges of 16-bit numbers map to which 8-bit numbers). The two tables are only slightly different. Most central office switches convert between the two companding laws. But if a switch neglects to perform the conversion, voice transmission still takes place in an understandable way—you might not even notice if speaking with a stranger.

Digital To Analog Conversion

At the receiving end, the 64 kbit/s digital stream is divided into 8-bit words, which are decoded to values for analog output. Here is where the two tables, for A or mu law, will show up again. Each received byte, every $1/8000$ sec, sets the value of the analog output. As closely as can be measured by the non-linear rule, the heights of these received values are the same as the pulses created by the sending Pulse Amplitude Modulator (PAM).

A low pass interpolation filter (Fig. 13) smooths the output into a continuously varying analog signal very similar (to our ears) to the input at the sending end. In effect the board fence is being reconstructed by first erecting every tenth board, to measure, then filling in the spaces between to match.

Compression vs. PCM

Note that each sample and each digital word in a PCM signal contains the absolute value of the analog input at the sampling time. Each measurement is independent. This completeness allows PCM to handle analog waves that jump from one extreme to the other while introducing a minimum amount of quantizing noise (Fig. 14). But this agility is called on only during the loudest passages of the highest frequencies, for example a high speed modem signal.

Most of the information in a human voice is conveyed by voice frequencies around 1,000 Hz, the middle octave of the voice range. At that frequency the voltage changes slowly compared to the 8000/second sampling rate. There is a strong correlation between adjacent samples in normal conversations. This means PCM contains more information than absolutely required to describe the voice sound. It is this extra information or redundancy that is removed by voice compression.

Compressed voice encoding finds ways to convey the voice sound without repeating information unnecessarily. Compression removes some of the redundancy by relying on the known patterns and statistical properties of normal speech, and perhaps the response of the human ear.

Applying assumptions that are valid for speech to the different situation of modem signals usually doesn't work as well because modem noise doesn't have the same properties as voice. To handle facsimile efficiently, a transmission system must recognize the fax signal and act like a modem to convert (demodulate) information back to digital form.

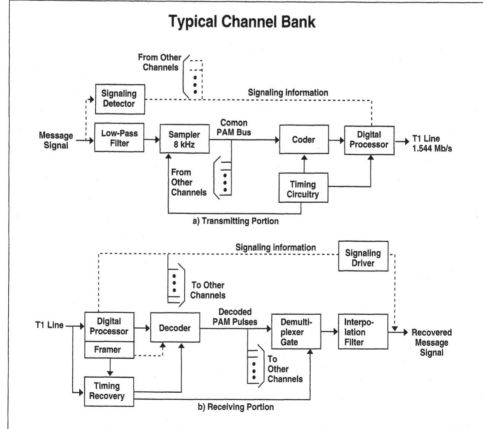

Fig. 13. Channel Bank block diagram shows separation of PAM and digital coding processes. Sending and receiving involve independent sections.

Differential PCM

Because voice signals change relatively slowly between successive samples, the change from one sample to the next (the differential) usually is much less than the full volume range. This means, for the same accuracy, fewer bits are needed to describe the change (the differential) than to describe the actual or absolute signal level anywhere in the full range (the way PCM does).

To improve the resolution provided by the reduced number of bits, DPCM contains a predictor. After each measurement, the sender calculates a "guess" for what the next sample is going to be, based on the signals already sent to the receiver. It is something like a moving average of the digital information. Rather than quantizing the next speech sample as a change from the preceding sample, the encoder quantizes the difference between the new sample of speech and what the predictor "guessed" it would be.

The receiver uses the same prediction process as the sender, so it comes up with the same "guess" for the next sample (not yet received). When the encoded information for the next sample arrives, it is added to (or subtracted from) the predicted value to get the actual digital value that is converted back to analog audio. The prediction process reduces the average differential that must be encoded, thus increasing the precision of the measurement for any given number of bits used.

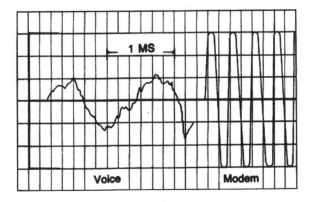

Modem sound Unlike Voice

1 MS

Voice   Modem

Fig. 14. Human voices show the largest component frequency at or near 1,000 Hz. A modem signal, to take fullest advantage of the channel capacity, will more often use the combination of maximum frequency and full amplitude.

Note that the receiving end must remember where the current level is at all times (the function of the predictor). Due to quantizing noise, this level may drift over time. An error on the transmission line can send it far off. To return the sender and receiver to the same levels, they are adjusted to zero each time the changes or differentials are zero for many samples. This happens when a speaker stops talking, which occurs frequently in normal conversation.

The technique of Differential PCM (Fig. 15) can work with any number of bits, but generally employs 4 bits per sample. There is no free lunch, however. On those occasions when the input signal really does change drastically between samples, the DPCM technique is not able to follow the input. The discrepancy introduces large amounts of quantizing noise and distortion. Consequently, 4-bit DPCM probably won't pass a modem signal of 4,800 bit/s.

Adaptive DPCM

DPCM was improved without increasing the bandwidth by an algorithm that assigns different meanings to the 4-bit digital words to suit different conditions. Specifically, the volume range represented by the 4-bit word increases or decreases.

For example, if successive voice samples suddenly were very wide apart, the differential normally expressed in four bits could not match the actual change. The ADPCM algorithm adapts in the situation of loud volume

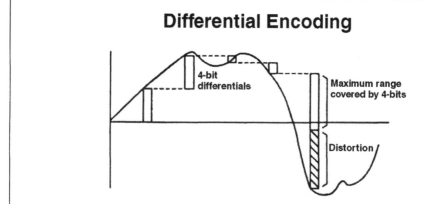

Differential Encoding

Fig. 15. Differential PCM transmits only the change since the last sample. By contrast, PCM send the absolute value of the analog input. Normal voice signals work well, as they change relatively slowly. However, sharp changes (high frequency and full volume) may exceed capacity and cause distortion

by increasing the range represented by four bits (Fig. 16). In the extreme, 4 bits might cover the range normally represented by 8 bits. This is the adaptive function.

While reducing quantizing noise for large signals, an adaptation to loud sounds will increase quantizing noise for normal or small signals. So when the volume drops, the ADPCM system reduces the volume range covered by the 4-bit signal. In this way the differential depends on the recent history of the analog input signal: DPCM becomes "adaptive," or ADPCM.

There are several types of ADPCM, including recommendations adopted by the CCITT like G.721 (in two incompatible versions, 1984 and 1988). ADPCM normally encodes speech at 32,000 bits per second. When a network is congested, an operator may want to reduce the bit rate per channel. Extensions to G.721 (in G.723, 1988) define 24 and 40 kbit/s encoding as well (3 and 5 bits per sample). A later version of ADPCM defined in 1990 (G.726) is able to adjust the bit rate by changing the number of bits per sample from 5 down to 2, or a minimum of 16 kbit/s. However, G.726 does not easily change speed "on the fly."

These are not to be confused with G.723.1 which is a new dual-rate scheme for low bit-rate voice (LBRV).

Each encoding speed for ADPCM is defined by a table of 31, 15, 7, or 4 entries. The table maps the value of the differential (between the measured signal sample and the prediction, after adjustment for the adaptive scaling factor) to a 5, 4, 3, or 2 bit binary number. It is the sequence of these numbers that are transmitted between speaker and listener. When received, the

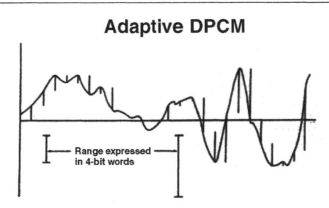

Adaptive DPCM

Range expressed in 4-bit words

Fig. 16. The range or maximum size of the differential can be changed to accommodate the input signal. Loud passages at high frequency increase the differential. A running average of the most recent differentials controls the range.

code bits are converted to a "volume" which is adjusted by the locally calculated adaptive scaling factor.

Bandwidth for signaling (dialing, off hook request for service, etc.) is a separate consideration.

ADPCM voice quality at 32 kbit/s is very close to 64,000 bit/s PCM. But 32 kbit/s ADPCM cannot handle modem signals at 4800 bit/s and faster as well as PCM does, hence the faster 40 kbit/s option.

Embedded ADPCM

The ADPCM algorithms described above use all of the encoding bits in the predictor circuit. That is, if the coding rate is 32 kbit/s, all 4 bits are considered when making the prediction of the next voice sample. The implication is that all of the bits must reach the receiver in order to keep the two predictors synchronized.

If a change in the encoding rate is desired (up for a modem, down in response to congestion), the end points must negotiate the change. This is not usually supported, so new calls may be rejected or blocked for lack of bandwidth.

To provide more flexible bandwidth allocation, "embedded ADPCM" was defined in G.727. The differences are:

1. The encoded bits are divided into 'core' and 'enhancement' bits. Think of them as "more significant" and "less significant." For example, 2 bits might be designated core, and 0 to 3 bits used for refinement of the signal (reduction in quantizing noise). Only the core bits are used by the predictors at the sender and receiver.

2. The output of the encoder is easily packetized.

In other words, the differential that is encoded is the actual signal minus the prediction based on only the core bits. If enhancement bits are not delivered to the receiver, but only the core bits, the two predictors remain synchronized. The sound quality during periods when fewer bits are being delivered is not as good as when all the bits are received, but the system continues to work smoothly.

The term 'embedded' refers to the fact that the quantizing decision tables for lower bit rates are subsets of, or embedded in, the table for the highest rate (40 kbit/s or 5-bit encoding).

Note that the enhancement bits are "optional" at the receiver. They are used if received, but their absence causes no problem. The transmitter could decide not to send the enhancement bits to make room for more connections over an access link with limited bandwidth. The network could discard the enhancement bits, in response to congestion, but maintain the connection if it delivers the core bits. The speed change is handled gracefully by the receiver, without audible disruption.

In a frame relay network, congestion at a switch is reported to both ends of the connection. The BECN and FECN bits are set in frame headers to give backward and forward explicit congestion notification (Fig. 17). These bits are always zero when the frame originates.

Frame relay terminal equipment is supposed to reduce the amount of traffic presented to the network while BECN and FECN are set by a congested switch. One way a VoFR FRAD could reduce traffic is to drop the enhancement bits on EADPCM connections until BECN is cleared (returns to zero).

To facilitate dropping only the less important bits, in case of congestion, one proposal put the core and enhancement bits in separate frame relay frames. The frame with the enhancement bits would be marked as lower priority by setting the Discard Eligible bit in the frame header. This approach is captured in Annex G of the VoFR IA.

The ITU recommendation for packetized voice puts the different kinds of bits in different fields in the same frame. This approach requires significant changes in the switches that make up public networks if they are to discard only part of a frame. Present frame relay switches do not look inside frames, so they cannot discard only the enhancement bits: a switch must discard whole frames. However, to minimize the number of frames generated,

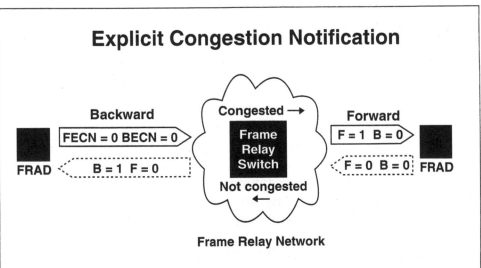

Explicit Congestion Notification

Frame Relay Network

Fig. 17. The frame relay header contains two ECN bits for forward and backward notification. Both bits start as 0; only a congested switch can set them to 1. Which bit is set depends on the direction of the frame with respect to the direction of congestion.

and the processing performed by the VFRAD, the VoFR IA stipulates the single-frame format of Annex F for compliant ADPCM implementations

Sub-band Coding

In this method, the speech signal is divided into a number of different frequency bands (five in one case) using a special type of filter called a Quadrature Mirror Filter. The signal in each frequency band is then quantized using a method similar to ADPCM. Since the energy in each band varies as the speech signal changes, the available bits are allocated adaptively to the frequency bands based on their energy content.

Adaptation minimizes the quantizing noise, but because each band adapts independently the total number of bits per second may vary from less than 10 kbit/s to more than 20 kbit/s. SBC won't fit neatly in a fixed-bandwidth channel, but it works fine over packet circuits or when used to record voice on disk for message storage, etc.

The speech quality produced by sub-band coding is good at rates as low as 16,000 bit/s, but these coders can put lots of delay into the system. You normally don't notice the delay since it's smaller than the delay on satellite circuits. But delay can affect the perceived voice quality and require echo cancellers. Of course if the audio is being digitized for recording, like voice mail, the delay is irrelevant.

Linear Predictive Coding

Linear predictive coding (LPC) compresses a voice signal to 2,400 bit/s and even less. LPC uses a far more complex prediction algorithm, to reduce the average differential that needs to be encoded. It requires several times as much circuitry, and used to cost 100 times as much as PCM.

Perceived quality is quite low—a Donald Duck impersonator is a common reaction. The voice of a familiar person may be recognizable, but barely so. The sound is intelligible, however, and is adequate for conveying instructions. Because of the high cost, LPC was found most often where the bandwidth was extremely expensive, as on early satellite circuits or leased land lines to remote areas. LPC has been largely replaced by better encoding schemes.

Code Excited Linear Prediction

CELP algorithms have been the rising stars for voice compression in the 1990s. There are at least as many variants as in ADPCM.

The essence of CELP is a code book, kept by the sender and receiver. Each entry in the code book itself is a list of numbers that represents a series of voice samples, not just one sample as in PCM. In fact, a code book entry could look a lot like a 10 ms sequence of 80 PCM samples. The resolution or the number of bits per sample is not important for this explanation.

If the numbers in the code book entry are plotted, the resulting graph will have a particular shape (Fig. 18). Each shape is called a 'candidate excitation vector.' A mathematical vector is a string of numbers where their order has significance. Here the numbers are the PCM values that describe the shape. The amplitude or height is for a typical volume of speaking voice.

To encode an analog voice signal, the CELP circuit operates on 10 ms blocks. First the encoder collects 80 PCM samples (linear values, before companding). The volume level is adjusted to "normal" to get a standardized "shape." Then that shape is compared with the "candidate shapes" in the code book to pick the closest match. The only information that must be sent to the far end is the index number of the selected shape and a loudness level.

At 10 ms per block, or 1/100 second, the 16,000 bit/s output allows 160 bits per block. That is a very large number and offers a lot of options in picking a vector. But then a lot can happen to a voice signal in 10 ms.

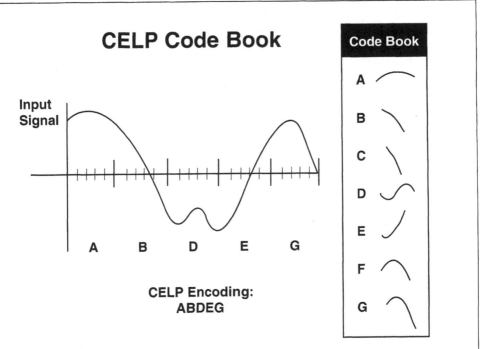

Fig. 18. The "code excited" part of CELP refers to a code book or dictionary of curve segments (vectors) that are selected to approximate the input signal. It is not necessary to send all the details of the shape (as PCM does), just the sequence of codes representing the curve segments.

Technically, the method used is called 'analysis by synthesis.' The best match is found by replicating the receiver's function of building up the original signal from a vector. The DSP tries to reproduce the input from several vectors, then picks the one that allows the reproduction to come closest to the original. That 'trial' reproduction is what the receiver creates by using the same code book entry (vector) and loudness factor.

Carefully selected code book entries can represent almost any wave form generated by a human voice. With relatively few code book entries (even several thousand is 'few' for well organized digital processing) the job can be done in real time by inexpensive digital signal processors (DSPs).

Unfortunately, modems and facsimile machines aren't voice so they are not usually encoded accurately. CELP systems often demodulate modem "noise" back to a digital data stream that can be packetized and intermixed with voice frames.

Low-Delay CELP

G.728 defines a version of CELP encoded at 16,000 bit/s and optimized for low delay (LD-CELP). The goal (achieved at a MOS score of 4.0) was to offer the same quality as ADPCM at 32 kbit/s. For lower quality, encoding bit rates down to 12.8 kbit/s are available.

To reduce delay, only 5 samples are accumulated in a block to determine a shape, which takes 0.625 ms rather than the 10 ms for basic CELP. Shorter accumulation times reduce delay, and the shorter blocks are quicker to process. Samples are linear pulse code modulation, taken at 8000 per second, which is exactly the first step of standard PCM before companding. Dividing 16,000 by 8000 gives two bits per sample; five samples per block means that 10 bits are available to encode each section or shape, that is, in up to 1024 different ways.

In practice, 7 bits are devoted to 128 idealized shapes (vectors). The other 3 bits indicate amplitude or volume. That allows the shapes to be stretched to fit the input signal. Knowing that an analog input may have infinite variety, even 1024 possibilities seems weak, and they would be if they were static.

CELP's well-deserved reputation for complexity comes from constantly updating the code book and filters, based on recent past experience. This is roughly similar to the adaptation of ADPCM, but in a way that requires far more computation. The amplitude factor is adjusted independently.

Human hearing enters the process as a "perceptual weighting" applied to the comparisons made between analog input and code book entries. How close the encoding comes to the original depends on how it will sound to a person. For modem signals, the perceptual part may be turned off to allow transmission of up to 2400 bit/s treated like voice.

Algebraic CELP

Another factor-of-2 improvement in bandwidth is achieved by ACELP: good-as-ADPCM on only 8 kbit/s. The difference is in the code book entries. Each is an algebraic expression rather than a series of numbers. DSPs manipulate the shapes more easily when they are encoded as mathematical functions rather than arbitrary vectors (lists of numbers).

By design, the amount of processing has been minimized. This approach reduces the processing time, which helps minimize overall delay.

ACELP as defined by ITU G.729 has been accepted by the technical committee of the FR Forum as one of two minimum interoperability algorithms for VoFR. It also has proprietary implementations.

Self-Excited Vocoder

'Vocoder' is a general term for any voice encoding device, and could be applied to any process described in this book. Self-excited refers very specifically to the way this design creates and updates its CELP 'candidate excitation vectors.'

Rather than have a fixed code book of shapes or vectors, SEV continually revises its code book based on the shapes of recent sample blocks of the analog input. At startup, there is a default list of random shapes that gets the process going.

With the code list constantly adjusted to the current traffic, SEV is claimed to have superior fidelity and adaptability. Good quality is claimed at encoding rates as low as 4800 bit/s.

Mixed-Excitation Linear Prediction

A further refinement in MELP introduces non-linear elements into the components of the code book. For example, rather than base the model on pure tones (periodic repetition) alone, MELP allows for non-periodic pulse trains. When starting or ending a speech burst, aperiodic pulses smooth the transition from silence to voiced sound and removes some forms of synthetic noise.

Dispersion of the frequency of the source in reproduced sound further smooths the sound as perceived by the listener.

MELP can mix excitation or source signals in different frequency bands. This technique reduces the perception of buzzes or unwanted tones in the reproduced speech.

An adaptive filter in the reproduction path improves the accuracy of the waveform, resulting in a more natural sound quality. This reduces the "Donald Duck" resemblance.

The Department of Defense chose MELP for its standard compression algorithm at 2400 bit/s. It scores far better than anything else they tested at that rate, and better than some methods running at double the bit rate (4800 bit/s). Processing requirements are not outrageous, and MELP performs well when background noise is present.

On the down side:

— compression does takes a fair amount of processing power.

— the resulting frame is 54 bits, which needs to be padded to a multiple of 8 in order to be encapsulated in frame relay. It won't fit in an ATM cell.

It is not expected that MELP will be popular in commercial applications.

Other Compression Methods

ITU's G.722 describes a way to transmit a 7 kHz audio bandwidth in a 64 kbit/s channel: the sampling is done at 16,000 times per second.

There also are many proprietary methods for low bit rate voice. In fact, at this time, all commercially available VoFR products are proprietary. This means they neither conform to the standards, nor interoperate with each other.

Questions of Voice Quality

When considering compressed voice, it may be important to examine several factors, in addition to the compression algorithm, that affect transmission quality. Some of them may also impact bandwidth requirements and therefore the overall design of the network, public or private.

There are at least two possible purposes for a digital voice channel:

1. To carry human conversations.

2. To carry modem signals, including facsimile.

Voice compression works most effectively on real voices. Modem signals are artificial replicas, using the same nominal bandwidth but in different ways. For example, to maximize the bits per second throughput, a modem will use all of the available frequency bandwidth and dynamic range (loudness) in a voice channel. Thus a modem will be far more demanding than the human voice.

To a human ear, a circuit may convey a spoken conversation with perfect clarity. The speaker's voice may be recognized easily. The voice quality is high. Yet a modem on that same channel will not be able to operate error-free at high speeds. The modem's stringent requirements for phase coherence and a need to send high frequencies at very loud levels with minimal distortion cannot be met with anything less than PCM.

Result: the "quality requirement" depends on the application.

Facsimile

Facsimile machines create an expectation that modem traffic will appear, with certainty, in both private and public voice networks. Where older forms of "digital" information were accommodated in digital form, the fax, with its internal V.29 modem, insists on working into an analog interface. Even with digital transmission equipment, the port on it will be analog.

An ADPCM channel cannot carry even a 9,600 bit/s modem signal, the standard speed for fax machines. Voice channels compressed to lower digital bandwidths, like 16 or 8 kbit/s, may not be able to handle the facsimile fallback speeds of 7200 and 4800 bit/s.

To handle fax, some networks assign certain voice channels to be full PCM at 64 kbit/s, ensuring fax machines work at full speed. At headquarters, the administration of many such channels may become a problem if it is necessary for the fax machines (via PBX programming, etc.) to find the PCM channels and avoid compressed channels. At remote sites, there may be only one or two voice ports on the frame relay access device (FRAD). With that few ports, it becomes seriously restrictive to provide separate voice and fax channels.

To simplify network administration, customers have demanded that compressed voice channels be transparent to facsimile. With voice compression hardware based on digital signal processors (DSPs, the same type of processor at the heart of a modem), the difference between a voice compressor and a fax modem is in software. When a voice compressor recognizes a fax connection (from its modem noise), it changes from voice software to fax software automatically. The transition between voice and fax is transparent to the end user, a PBX, or the fax machine.

In this way the analog to digital conversion into the network reduces the bandwidth requirement from 64,000 bit/s to the base speed of the fax modem. This usually is 9600 bit/s. The newer fax modem speed of 14,400 bit/s was not supported widely at this writing, forcing such transmissions to fall back to 9600. This may not be all bad.

One difficulty with 14.4 fax calls on compressed voice channels is that the guaranteed size of the network channel assigned for voice (TDM bandwidth or committed information rate for a frame relay virtual channel) may be 9600 or less. TDM networks would need to be reconfigured to increase the channel capacity to 14.4 kbit/s. Packet networks, including frame relay, can support speeds of 14.4 K, but part of the data stream might be discarded if it exceeds the CIR. Keeping the fax speed at or below the compressed voice bit rate ensures that the fax transmission has the same protection against discard as the voice—usually high priority transmission from the FRAD and a guarantee based on CIR in the network.

Lost Frames

For data protocols, loss of a frame triggers an automatic retransmission of the frame. The error is corrected.

Voice connections can't wait for a second try at a lost frame. The conversation just moves on. There is no practical way to correct errors in real-time voice at the transmission level.

Fortunately, the frame loss rate for Frame Relay Service is designed to be small, and in fact is small in practice. The size of each frame is also small, typically representing a 20 to 50 ms interval of speech. If a 20 ms segment of speech disappeared, it could produce a discontinuity in the received sound, but within a syllable; 20 or 50 ms is not long enough to lose an entire word. Then again, if a frame loss occurs in a vocalized pause ("ahhhhhh") it might not be noticed at all.

In the worst case, when an error burst triggers the discard of many consecutive voice frame, we can always use the "Huh?" protocol. That is, the listener asks the speaker to repeat, the same as if the listener were distracted or there were a loud noise that drowned out the speaker. Any user of cell phones would be happy to have so little difficulty (compared to periods of 10-20 seconds of static as a car moves between cells).

Digital Speech Interpolation

A characteristic of normal voice conversations is pauses. Usually one party listens while the other speaks; at times, both are silent; in only a few cultures do both sides commonly talk at once. When voice is sent as frames, it is possible to use something like statistical multiplexing to eliminate these pauses from the transmitted voice signal. DSI reduces the average bandwidth needed by an individual voice path on a multi-channel trunk. The instantaneous bandwidth requirement alternates between the encoding rate while speaking (e.g., 8 kbit/s for CELP) and zero during silent periods.

Because DSI relies on statistical probabilities, it works only when there are many voice channels. In the old analog days it took 72 channels to get an additional 1.5:1 compression; for a 2:1 gain, about 96 voice connections were needed. Newer digital implementations, using faster switching, claim nearly 2:1 compression with less than 24 active voice channels.

Hardware and software digitize and compress the audio input continuously, even when the speaker is silent. But the processor marks those frames that contain nothing above a set threshold of loudness. These frames of "silence" may be discarded by the sending FRAD. A good strategy might continue to send silent frames for a short time, from a fraction of a second to several seconds, in case the speaker picks up again. In that way the background noise level is consistent for the listener.

When the sender decides to stop sending frames because the speaker is silent, the background noise must be recreated by the receiver. "Synthetic" background may be simple white noise (static), or it could be repeats of some recent "frames of silence" received at the end of a speech segment.

A possible penalty of DSI is some 'clipping' or loss of voice information. If all speakers talk at once there may not be sufficient instantaneous bandwidth for them all—some frames may be discarded due to congestion. After a silent interval, the sending FRAD may miss the first frame (or several) if the encoder is not quick to recognize sound above the loudness threshold. There is also some portion of the very beginning of an utterance that will not exceed the threshold, leading to abrupt starts for speech segments.

It is up to the designer of the voice compression and transmission equipment to find strategies that minimize clipping. These could include, for example, keeping the most recent frame in memory, even if "silent," and sending it before the next voice frame if that next one is seen to be active.

DSI also requires considerable overhead to let the device at one end tell the other end which conversation is currently being transmitted on each channel.

Recovery by the Algorithm

The stream of frames at the receiver can be interrupted because of frame loss in the network or silence suppression at the sender. For whatever reason, the receiver has to restart when the frames resume. For PCM, where every byte is a complete and independent measurement, there is no problem with interpreting the information.

But what happens when the algorithm includes a predictor that relies on recent valid frames? What about a CELP algorithm that updates its dictionary of symbols from voice samples?

Lost frames may mean the receiver's version of the predictor or the dictionary can be different from the sender's. It should be clear that arriving information, interpreted on the basis of outdated values at the receiver, will not produce the expected results in terms of accurate reproduction of the sender's voice.

Recovery depends on the two ends getting synchronized again, a process called 'convergence.' Usually it is the receiver that has to adjust, the faster the better.

ADPCM is fairly good at recovery. Its predictor depends on only a few samples, which are taken 8000 times per second. The time lapse is small, and generally ignorable. Certain LPC algorithms, which depend far more on the predictor, can take seconds to converge, making the frame loss very noticeable.

Most of the latest algorithms, like CELP and MELP, are designed to recover quickly. Reasonably accurate reproduction should be available in 10 to 20 ms after a lost frame; that is, perhaps within the time of one compressed voice frame.

For comparison, a channel bank is allowed 50 ms to synchronize with its T-1 line after an error or a 'loss of framing' event. To most telephone users, the loss and recovery sound like a brief click.

Most observers agree that recent packetized voice techniques provide more than adequate quality and quick recovery from the loss of an occasional frame. A high rate of frame loss will produce unacceptable sound quality, but such a situation indicates the need to repair or redesign the frame switching network.

Chapter 3

Voice Over Frame Relay

Up until the 1990s, frames used for voice were generated at a constant rate to fill a TDM channel. The contents of the frames represented another emulation of the copper loop (see Chapter 1).

Emulation is how an ATM network provides voice transport: 'circuit emulation service.' CES is a way to carry PCM encoded voice (in a T-1 or E-1 format, or a DS-0 of 64 kbit/s) on a stream of regularly repeating ATM cells. The cells are generated at all times that the TDM circuit is "up," even if all the voice circuits on it are idle. CES is more TDM emulation than voice over ATM.

The Frame Relay Forum Technical Committee (FRF-TC) has worked for several years on an implementation agreement (IA) defining "true" packetized voice. The model design transmits frames only when a speaker is actually talking. It allows for idle connections, and silence suppression within a call, by not sending frames when there is no speech. This approach greatly improves bandwidth efficiency. The IA defines a 'SID' frame type (Silence Information Descriptor) to indicate the start of silence, but leaves to each vendor the decision on what 'silence' is.

The IA is presented in Appendix C. We will refer to it at many points in this chapter as we explain its provisions and procedures.

This chapter does not apply to "unpacketized" voice on TDM or circuit switched connections. Information passes transparently through TDM circuits; the network doesn't examine the information, but passes it blindly between locations. Therefore, a path and bandwidth must be reserved for each connection: users cannot share network resources and the network cannot provide statistical multiplexing or traffic concentration.

For bandwidth efficiency and low cost, as well as availability into the 21st century, frame relay is the optimal packetized network to carry voice.

Low delay is a highly desirable feature for voice. Because FR networks don't correct errors, the switches process frames quickly and introduce minimal delay (about 5 ms per switch).

FR is connection-oriented, which means the network assigns a fixed path to a logical connection. Having one path ensures that frames arrive in the order they are sent. The sender doesn't have to format the frames with sequence numbers (at least not to keep frame order); the receiver needn't check for frame sequence. Yet there are no permanently assigned resources along the path except just enough memory to identify it. All network users share the lines and frame switches.

Packetized voice is characterized by certain features, described in this chapter. All are considered important for voice over frame relay, and are generally implemented by vendors of Voice over Frame Relay (VoFR) equipment. If any of these elements is missing, there will be some potential for reduced quality.

VoFR IA Essentials

The requirements to achieve compliance with the FRF IA on voice over frame relay are limited to the compression algorithm, certain signaling transfer syntaxes, and the form of the frames. However, the frame format implies fragmentation of long data frames that share a 'data link connection' (DLC, also known as a 'permanent virtual circuit,' PVC) with the voice traffic.

In defining the requirements, the Technical Committee reasoned that ADPCM is capable of carrying fax and DTMF tones as compressed voice. Therefore there is no need for a separate transfer syntax for DTMF or fax (Annex A or D). On the other hand, CS-ACELP cannot maintain the low distortion required for DTMF, nor pass any fax signal. So its class also requires Annexes A and D.

Both classes need CAS (Annex B) for on- and off-hook designation. Since it is there, it may also be used for conveying dial pulses and E&M signaling.

Packetization

Of course packetization is essential, but not just any kind. To gain interoperability, the frame relay (FR) format is defined carefully (Fig. 19). The same frame header and trailer are used for voice and data. Packetized voice is sent as payload in the FR frames.

Blocks of digital information are generally called packets or frames. The term 'frame' will be preferred here, in the case of VoFR.

Frames can be separated from each other and still retain their identities. Excluded, therefore, is digital voice like pulse code modulation (see Chapter 2). PCM is sent as a constant byte stream in a time slot on a DS-1 line. Also excluded is digital voice sent in continuous frames that carry no channel identification; these are used in TDM transmission systems.

Frame Addressing

By definition, each frame or packet identifies the specific conversation (or conversations) it contains. The ID is called an address. The basic address is located in the frame header; the 'data link connection identifier' (DLCI) marks a 'Permanent Virtual Circuit' (PVC) or 'Data Link Connection' (DLC). The DLCI value has only local meaning; that is, between two adjacent FR devices that have been configured to agree which connection the PVC carries.

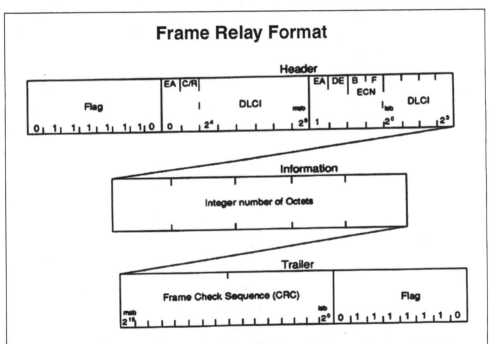

Fig. 19. Frame format for Voice is the same 2-byte header and 2-byte CRC as any data frame on a FR network. The C/R bit goes end to end between terminal equipment. The network indicates congestion by setting (=1) the Backward and Forward Explicit Congestion Notification bits. EA is 1 in the last byte of the header; the FRF also defines a 4-byte header. HDLC flags mark the beginning and end of a FR frame.

Public frame relay services and the switches in private frame relay networks deal only with the DLCI. Each DLCI is one PVC. The carrier will charge for each PVC provisioned, so users often want to share a single PVC among all the applications at a branch office. The VoFR IA defines how to mix voice and data using subframes and additional subframe addresses.

A PVC links two points; there is no broadcast or multidrop service at this time. Any FR frame with a specific DLCI always goes to the same receiving device.

VoFR Subframes

The VoFR address is extended with additional bits, the 'channel identification' (CID), in a subframe header (Fig. 20). The combination of these address numbers (DLCI and CID) identifies a connection on a physical link or port. The same pair of numbers on a different port in general will be a different connection (you could force them to be the same connection). In a sense, the port ID on any device is part of the address. Saying "CID 22 on DLCI 101" may be ambiguous if there is more than one port or TDM channel for frame relay traffic at that location.

Local area network (LAN) data frames also may have a subheader with an address, but usually it is called something else, like a higher-layer protocol address. For example, on the Internet, the Transmission Control Protocol (TCP) may carry Telnet, FTP, and HTTP protocol data units from one device that uses only one IP address. TCP, coming between IP and the other protocols, provides 'port' numbers to distinguish one connection from another.

Other protocols may use a subaddress called a Sub-Network Access Protocol. A 'SNAP header' also identifies individual logical connections when there are many sharing one physical connection or address.

Only one TCP port address may be encapsulated in each IP frame. In contrast, as indicated in the figure, VoFR allows 'packing' of many subframes in one FR frame.

Note that the last subframe header must not have a 'Payload Length' (PL) field; the 'Length Indicator' (LI) bit is 0. However, all the subframes that are "not last" must have LI=1 and a PL value (number of bytes, in binary) as the third byte of the subframe. Explicit PL values allow the receiver to separate the subframes. The assumption is that the last subframe is whatever is left in the frame, whose end (octet N) is indicated by the trailing HDLC flag, minus the 2-byte CRC.

To understand the reasons behind the VoFR frame format and addressing, some background is necessary.

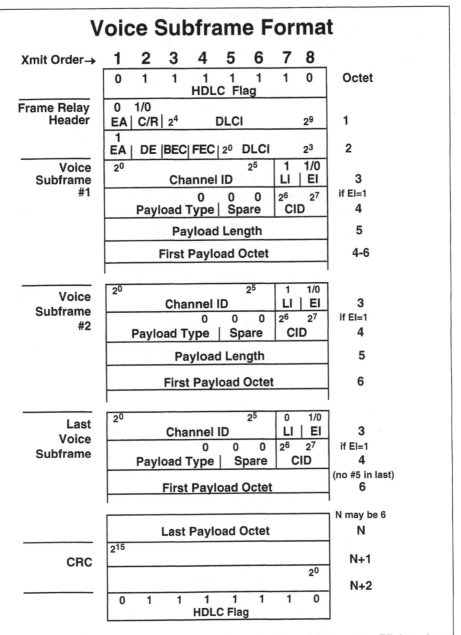

Voice Subframe Format

Fig. 20. Frame format for Voice over Frame Relay, shows the FR header (2 bytes) and the subframe headers that follow. The network reads only the FR header; the customer's equipment uses the subframe header to distinguish among different conversations and/or data streams. The subframe header address may be viewed as an extension of the FR header address. The effect is to create many logical channels within one PVC.

For the G.729 algorithm, a speaker causes one subframe to be created every 10 ms. Without packing ("packing" in VoFR means putting more than one subframe in a FR frame), that makes the load on the frame relay network 100 frames per second; four speakers could generate 400 frames per second. By contrast, a file transfer process may generate less than 5 frames per second onto a 56 kbit/s access line.

Frames per second (F/s) are important because the capacity of a frame switch more often is limited by its ability to process frames than by its bandwidth. Short frames can mean a switch will reach its F/s limit long before it approaches its throughput capacity (bits per second). With 30, 60, or more channels on a large voice device, one VFRAD could throw 5000 or more frames per second into a switch port, posing a serious threat of switch overload if many customers had the same idea.

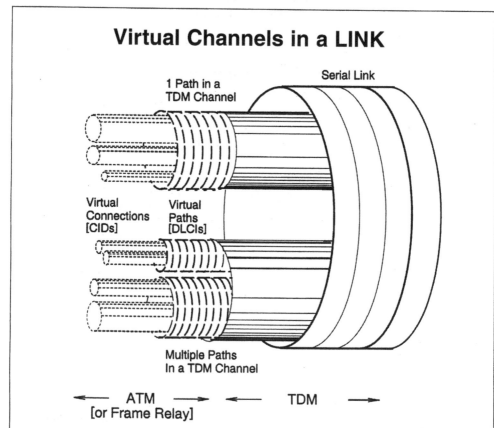

Virtual Channels in a LINK

Fig. 21. The VoFR Channel ID corresponds to the virtual channel identifier in ATM. In parallel, the frame relay DLCI corresponds to the ATM virtual path identifier. Both parts of the address and the port (or TDM channel) are needed to specify a connection uniquely.

The voice-equipped FRAD (VFRAD) puts part of the address into a sub-frame header as 'channel identification.' CID can multiplex many voice connections over one network connection (DLCI or PVC). Perhaps more important to network operators, VFRADs also offer a way to let many conversations share one FR frame. Sharing, also called 'packing,' dramatically reduces the number of frames sent.

In the example of 4 speakers, submultiplexing reduces the frames per second to 100, the same as for a single speaker, if each frame contains one subframe from each call. There is no reason not to put more than one subframe from a call into the same frame relay frame. A number suggested as a good trade-off between latency and F/s is 5 subframes from each channel. Packing 5 subframes of G.729 encoding in each frame brings the frames per second down to 20–a much more manageable number than 400.

Packing 5 subframes from 255 channels isn't practical–the frame size is over 13,000 bytes. But then, if there is that much information, a decent frame size results from packing only 1 subframe each from half the channels in each FR frame.

Those familiar with cell addresses in ATM may notice a parallel here. The ATM 'virtual path identifier' or most significant portion of the address corresponds to the frame relay DLCI. A second cell address field, the 'virtual connection identifier,' distinguishes individual sessions or logical circuits that share a virtual path (Fig. 21), just like the VoFR channel ID. Again, it is the combination of VPI (DLCI), VCI (CID), and port that identifies a specific connection.

The original intention for the frame relay service was to use the Data Link Connection Identifier (DLCI) to mark one logical connection; for example, one data connection. As demonstrated above, applying the same rule to voice conversations leads to unacceptably large numbers of voice frames.

From a high-level viewpoint, adding voice will introduce significant traffic, measured in frames per second. However the VoFR technology scales up well; adding more voice channels (CIDs) does not increase the frames/second load. Mixing data with voice on the same PVC should keep the number of frames near the number generated by voice alone.

The VoFR IA imposes no restrictions on how submuxing is done. A VoFR frame might contain only one voice, data, signaling, or fax subframe. The Payload Type (PT) in the subframe header will indicate which it is (Fig. 22).

Practical limitations may apply. For example, the default in frame relay for maximum frame size is 1600 bytes. That is big enough to hold 122 CS-ACELP subframes, if not 255. Longer FR frames are possible, when both sides of the interface agree, but in frames over 4000 bytes the 16-bit CRC cannot catch all combinations of bit errors in the payload.

All proprietary VoFR equipment has similar submultiplexing to share PVCs among many connections. Submultiplexing is attractive because carriers charge for each additional PVC (DLCI). Carriers are certain to charge for each Switched Virtual Circuit when SVC service is offered.

The small subframe size in the VoFR IA for G.729 CS-ACELP could offer an easy migration path to ATM networks. The ATM Forum could, for example, adopt the VoFR compression algorithm and subframe header format. An ATM cell with a 48-byte payload might carry only one voice subframe, with one conversation per ATM virtual connection. This approach has a cost in low bandwidth efficiency from cell header overhead.

There is room to pack three G.729 subframes into a cell, with room for an ATM convergence sublayer (subframing format). The additional 20 ms of accumulation delay (three 10 ms intervals instead of one) would be offset by a reduction in transmission time of almost 7 ms. This comparison is between frame relay on 56 K (3 x 2.3 ms) and ATM service at 1.5 Mbit/s (1/30 ms).

Voice Compression

While the VoFR implementation agreement recognizes it is possible to put PCM-encoded voice into frames, it is not very practical. PCM bits alone need 64 kbit/s, a full DS-0 channel or time slot. Adding the overhead of frame and subframe headers and error check code means using more bandwidth than required for the best 'toll quality' service today. Except for compatibility during a transition period, there seems little incentive to use anything but a TDM channel for PCM-encoded voice.

Applying silence suppression or voice activated transmission (VOX) to PCM frames is easy. The receiver recovers instantly–it uses only current

Payload Type in Subframe Header

Bit	1	2	3	4	
	0	0	0	0	voice transfer
	1	0	0	0	dial digits
	0	1	0	0	signaling bits; or silence insertion descriptior (PCM)
	1	1	0	0	Fax relay
	0	0	1	0	Silence Indication; any transfer syntax

Fig. 22. All other values of the Payload Type field are reserved.

information and has no predictor to update. But VOX cannot be relied on to keep the average bit rate, including frame header overhead, below 64 K. This is obvious if you consider fax or modem users.

How Much? = Which Algorithm?

The only real question is how much to compress. Advances in DSP hardware and clever software give 32K ADPCM quality on only 8 kbit/s. The extra cost of hardware is recovered quickly in reduced line costs.

ADPCM is the first step, in terms of bandwidth saved: half, or 2:1 compression. It is also cheap to implement and old enough to be stable. But it is not easily packetized in its original form. Later versions are packetized, and widely deployed (for example, the now discontinued IACS product from AT&T). More processing (CELP or another compression algorithm) can reduce the bit rate representing the voice sound to 1/8 that of PCM, about 8,000 bit/s compared to 64 kbit/s. Proprietary methods encode at 5.3, 4.8, or even 2.4 kbit/s.

Late in 1996, the FR Forum changed the default compression algorithm for VoFR from ADPCM to Conjugate Structure Algebraic Code Excited Linear Predictive (CS-ACELP) coding defined in G.729 from the ITU. Later, it put G.727 ADPCM back in as one of two compliance classes. The G.729 algorithm (Class 2 compliance) is appealing for its low bandwidth (8 kbit/s), relatively low processing requirements, and for the worldwide acceptance expected because it has been adopted by the ITU. Annex A to G.729 describes a simpler encoding algorithm, compatible with the original decompression process of G.729, that takes still less processing. It appears favored for implementation. G.729, like other low bit rate alogorithms, does have a problem in terms of bandwidth efficiency.

There is a trade-off between low delay and high bandwidth efficiency. Low delay means smaller frame payloads, but the frame headers stay the same size. Therefore, overhead increases as a percentage of transmitted bytes when the delay drops. Two common examples:

– A minimum-size 16-byte voice frame (37% overhead) will cross a 56 kbit/s interface in 2.3 ms.

– A 1500-byte Ethernet frame (<3% overhead) takes 214 ms/frame.

An interval of speech must be accumulated in digital form, then compressed by a computer process before it can be sent. G.729 voice frames are kept short (10 ms) to minimize this 'accumulation delay.' Typically a voice frame represents no more than 50 ms of speech.

After 8:1 compression, 10 ms of sound becomes just 80 bits, not counting headers. A complete frame consists of those 10 bytes plus overhead. Not packing–putting only one subframe per frame–needs a total of 6 bytes of header overhead, admittedly not very efficient at 37% overhead. Additional subframes (10 payload + 3 header) are 23% overhead, so packing multiple subframes into a frame brings down the average overhead and improves efficiency.

When G.729 was adopted, the question of cost remained. At least four organizations claim ownership of patents covering some features of the algorithm. Early indications are that per-channel license fees will add $500 or more to the price of a 30-channel device. However, variants in the algorithm not covered by the patents are available and will likely be offered to users who want to reduce costs and don't mind giving up strict compliance with the IA.

The only drawback will be the need to buy all packetized voice equipment from one vendor. Not too much of a problem: this is exactly what has happened so far.

If the license is too expensive, the FRF TC will find another algorithm, in addition to Embedded ADPCM. Some one will get it standardized and listed in an annex to the IA.

Transfer Syntax

Both ends of the voice connection need to agree on how they are transferring information. Since there are no provisions for automatic negotiations by FRADs, they need to be configured for voice as they are installed. There are options allowed in the IA. Each end of a channel (CID as well as DLCI) must be specified the same.

Part of the setup for each voice algorithm defines how the compressed voice bits should be carried in the payload of the subframe. Since each method produces different bits, they are arranged differently inside the subframe.

While drafting the VoFR IA, the FRF received several contributions specific to one or another compression algorithm. Almost all contributions were included in the IA. They are annexes, to make it easier to add and drop algorithms in the future.

Dealing with the details of the syntaxes is necessary only for equipment designers or testers. End users will configure VFRADs by selecting the encoding speed or the algorithm by name. You will never need to set up the framing for the transfer syntax.

All transfer syntaxes share two signaling syntaxes and one fax relay syntax, defined in separate annexes. Again, the details will be handled by the equipment maker. You will specify the type of signaling, for instance, as DTMF (tone) or pulse (rotary).

In use, the alternate transfer syntaxes may be substituted on a channel (CID within a DLCI) at any moment, to transfer signaling or relay a fax image instead of sending voice. To distinguish these functions, the sub-frame header has the 4-bit 'payload type' (PT) field.

There are primary payload types, all indicated by the PT field being all zeroes, and signaled payload types, where the PT field contains a specific non-zero value. The primary payload is configured at installation. When the channel carries anything else, the payload is signaled with PT.

A typical application is a tie line between a branch office and headquarters. A fax machine at the branch dials up HQ. The FRAD, configured for voice, sees the dialing fax machine as a voice call and connects to the PBX at HQ. When the fax machines issue their modem tones, the VFRADs recognize a fax call. As they switch software to become fax modems, they change PT to 'facsimile relay' as the subframes begin to carry fax information. How the VFRADs originate and use the PT=fax indication is up to individual vendors, within the fax specifications (T.30 and T.4) and the procedure outlined in the IA.

In some organizations, fax machines have dedicated trunks, which never carry voice. In these situations the VFRAD could be configured for facsimile relay as the primary payload type. Since there is no way to signal a change to voice, these trunks would be limited to fax at all times.

Data Transfer and Fragmentation

The most urgent need for fragmentation comes where voice and data share a slow (56 or 64 kbit/s) access line to the FR network. At that bit rate a 1500 byte Ethernet frame inserted between two successive voice frames will delay the second voice frame almost 1/4 second. This is unacceptably long. Moreover, it will be variable, creating large jitter.

When a FRAD is handling voice, or exchanges data with one that does, the expectation is that data frames will not exceed some relatively short length, configurable to something like 128 or 255 bytes. This reduces jitter in the delay to send a voice frame and in the time it is received.

The VoFR IA doesn't demand that data frames be fragmented when voice and data are on different PVCs. It just makes good sense to minimize jitter in this way. But if a data connection shares a PVC with voice (same DLCI), the data is placed into subframes according to the 'data transfer syntax' (Fig. 23).

A "data" CID must be configured for data, just as each voice CID must be configured for its voice parameters and syntax. That is the only way both ends know that the channel is for data. Note that the first four CIDs are reserved for a technical reason: those values (0-3) in that position after the frame relay header, are used to identify specific, common data protocols encapsulated according to another specification (FRF.3, which is based on RFC-1490).

Since the subframe's payload length field is only 8 bits, the payload can't exceed 255 bytes ("two to the eighth") including the 1-byte 'data syntax header.' Therefore longer data frames must be fragmented into pieces no larger than 254 bytes and placed into successive data subframes with the same CID.

The 2-byte VoFR data fragmentation header contains the 'Begin' and 'End' indicator fields for the first and last fragments of a larger data unit.

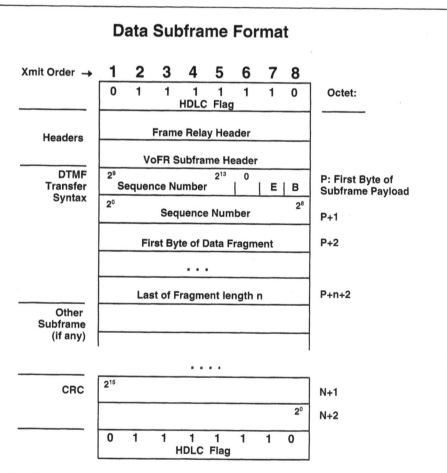

Fig. 23. Data transfer syntax requires a channel (CID) be configured for data. Each subframe with that CID will be treated as part of the specific data stream. The Payload Type is all 0s, primary payload type. The Beginning or End bit is set if the payload is the first or last fragment of a larger data frame (recall PL <=255). For a small, single fragment data frame, both B and E =1.

Subframes that are neither B nor E (both bits = 0) contain a middle fragment. If a full data frame fits in one VoFR subframe, both B and E are set = 1.

A 13-bit sequence number allows the receiver to detect a missing fragment. It increments by 1 for each subframe in the CID, and cycles back through zero.

Data fragmentation facilitates prioritization. Small fragments are fitted more easily into the outgoing queue between high priority voice frames.

The exact way to fragment data frames is the subject of another FRF IA, working title "FRF.frag" and pre-assigned FRF.12 as its number, when approved. It is expected to be published in 1997.

Fax Relay

The dominant type of facsimile machine is Group III ('group three'). These machines have analog interfaces that mimic a telephone. Fax machines plug into any standard phone line, direct from the central office or behind a PBX or key system. Fax can show up on any line a phone might use, including a VoFR path.

Fax modems emit a noise that is not at all like voice. Even the best voice compression algorithms perform poorly on these sounds. A fax machine trying to connect over a compressed channel may drop back in speed to 4800 or 2400 bit/s. It could fail to connect.

One solution might be to prevent fax calls from entering voice-only VoFR devices. An alternative PCM voice port might be time division multiplexed onto the same link, or faxes might be forced to use the public phone network. Many organizations put all their fax machines on trunk lines from a central office.

Either tactic significantly reduces the savings from VoFR.

That is why most VoFR equipment offers an option to handle fax on the same port with voice. This allows a PBX to "not care" if a caller is human or a fax machine when routing calls to a VoFR transmission device.

To handle a facsimile machine when it appears on the voice port, a VoFR device adjusts its DSP software to act as a fax modem. The audible fax signal is converted (demodulated) to a baseband digital signal:

– T.30 handshaking between images is V.21 modulation at 300 bit/s;

– T.4 image encoding can be anything from 14,400 bit/s (V.17 or V.33) down to 2400 bit/s (V.27ter).

Fax image bytes are sent as a series of FR frames, one frame every 40 ms. The size of the frame increases in proportion to the modulation speed. The transfer syntax for fax relay is Annex D of the VoFR IA.

To simplify the VFRAD, most of them today support 9,600 bit/s, V.29 modulation, the top speed of the most frequently installed Group III faxes.

Newer fax machines capable of 14,400 automatically negotiate the lower speed when the VFRAD won't go faster. The 14,400 speed will appear in VFRADs in the future. They too will automatically negotiate speed with fax machines and with VFRADs that have 9600 as a top speed.

Some networking considerations may limit the usefulness of 14,000 bit/s for fax.

When engineering a network for voice and fax, consider the committed information rate (CIR) to assign to the PVC and how each subframe channel in that PVC will share the CIR. Consider that faxes apply a constant carrier when transmitting, and offer no silent periods to save bandwidth via VOX or silence suppression.

So what should the CIR be, 8000 or 9600 or 14400? Higher CIR means higher cost. Some carriers offer CIR in multiples of 8000, or a power of 2 times 8000, or maybe multiples of 9600. Getting that extra 1600 CIR could be expensive. And if the number of voice channels fills an access line (4-6 on a 56K line), then increasing the CIR requires an additional line or a faster access speed--serious expense.

An alternative is to limit fax demodulation to 7200 or 9600 bit/s, to keep it within the CIR created for voice connections. If fax connections use no more bandwidth than voice connections, neither voice nor fax frames will be marked DE. This is good.

Actually, you can push the limit a little. The 8 K encoding rate does not include all the overhead, so the actual bit rate on a voice channel will be higher than 8000. The fax modem speed is 9600, with the only addition being the frame relay header. The two values may be close enough.

A voice algorithm that encodes at 8 kbit/s needs a CIR for its PVC of about 9600, to cover frame headers. If the fax speed allowed on that same subframe channel is 9600, any difference in performance shouldn't be significant. A fax speed of 14.4 kbit/s, however, may exceed the assigned CIR (will exceed, if there is only one voice channel on the PVC). If there are many voice channels within the PVC, then statistical averaging will reduce the problem of a fast fax channel.

Going above the CIR causes some frames to be marked 'discard eligible' (DE) by the network. Those frames face a higher risk of being dropped by a congested switch. How lost fax frames will be handled is up to each vendor, but the received fax will be missing a portion.

A fax machine when printing has no ability to start and stop the paper feed quickly. There are two consequences:

– Playback into the receiving fax should be continuous.

– Fax machines can't pause for error correction.

"Not pausing" is a fair match to a frame relay service, which doesn't retransmit lost frames. The 'forward error correction' (FEC) in the fax transmission protocol is there to correct minor bit errors from line hits; it cannot replace whole frames.

One way to obtain steady output is to increase the amount of the initial delay (see the section Delay Jitter later in this chapter). That ensures there is always a waiting subframe in the receive buffer.

While not part of the IA yet, some VFRADs 'spoof' the control sequences. This allows the calling and called fax machines to perform the necessary handshakes and negotiations on speed, resolution, page size, etc., with the VFRAD, not each other. Spoofing compensates for delay jitter in the FR backbone and can even correct incompatibilities between fax machines.

If the vendor is really serious about delivering fax accurately, the entire process could be spoofed, including image transfer. The sending machine would dump a page into the VFRAD. The two VFRADs would transfer that entire page image, retransmitting portions if necessary to correct errors or replace dropped frames. Only when a complete page had been received (with retransmissions if necessary) would the receiving VFRAD start "printing" to the destination fax machine.

Signaling Modes

Some of the longest discussions over the VoFR IA centered on signaling. This is a complex issue, not well understood by data-oriented people, nor by all voice-oriented people either. To help us all, Chapter 4 is devoted to the more common voice signaling methods.

To reach an agreement, the FRF-TC reduced coverage in the first version of the document to only the transport of signaling in its three essential forms.

DTMF Tones

To pass DTMF dialing tones as voice, VoFR devices must use an encoding algorithm with a very modest compression ratio. ADPCM at no better than 2:1 works. Algorithms with higher compression ratios can't reproduce the tones with adequate precision. The tones should not be compressed with any algorithm likely to be applied widely to voice.

When the algorithm for compatibility is CS-ACELP (G.729), DTMF dialing over the FR network will have to be handled as coded messages. The VFRAD's DSP (or DTMF detector chip) will recognize DTMF tones and convert them to digital symbols for transmission. The receiver's DSP can recreate the tone pairs to meet the precision specifications for these signals (frequency, loudness, and distortion are tightly controlled).

The DTMF transfer syntax (Annex A to the IA) can preserve the duration of a tone to within 1 ms of the time the caller pressed the key. The trick is to look at relatively short windows, 20 ms here. Whenever DTMF activity is detected, the VFRAD starts sending signaling subframes on that channel. Sequence numbers in the subframe header act as time stamps too (Fig. 24).

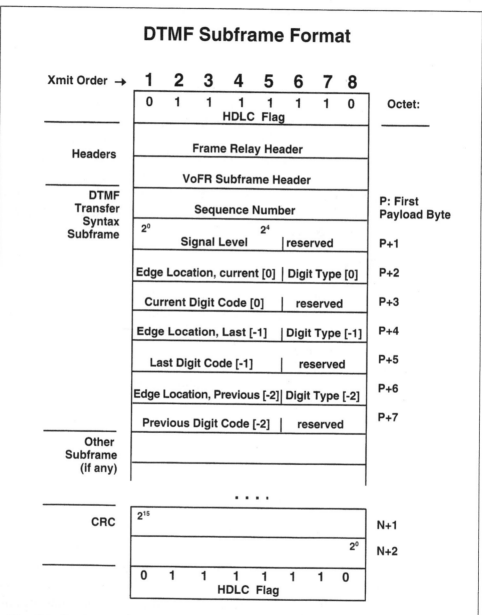

DTMF Subframe Format

Fig. 24. During DTMF activity, the VFRAD encodes the tones every 20 ms, indicating which tone and start, stop, or continue. The sequence number allows detection of lost frames and acts as a time stamp.

20 ms is too short a period for more than one valid DTMF event to happen. At most, only one valid tone pair can start or stop within a 20 ms period. According to the Bellcore rules, tones shorter than 23 ms should be rejected; receivers are not required to recognize "off" intervals shorter than 40 ms. The syntax looks for the presence of a tone or a transition, between on and off, then sends the necessary information in a signaling frame: the digit, on or off (Fig. 25), and the position of the transition within the 20 ms period (in ms after the start of the 20 ms period). If a tone

Codes In DTMF Transfer Syntax

Digit Type	Type Code
Digit off	000
DTMF	100
Reserved	all others

DTMF Digit	Digit Code
0	00000
1	10000
2	01000
3	11000
4	00100
5	10100
6	01100
7	11100
8	00010
9	10010
*	01010
#	11010
A	00110
B	10110
C	01110
D	11110

Fig. 25. Code values have been assigned to all 16 combinations of DTMF tones, though only 12 appear on most keypads. In future, other tone plans like SF or MF may be added to the Digit Types, requiring a new table of digit codes.

remains on from the preceding interval and throught the current interval, the transition location is 0.

Some autodialiers generate tone bursts that last only 10 ms, They seem to work, implying that tone receivers in CO switches are much better than the rules require.

For redundancy, each DTMF signaling subframe contains the same two bytes of information for three successive 20 ms periods: the current one first, then the two previous periods. Note that one subframe can contain two valid events in the three 20 ms periods: on/off or off/on. The power level field always refers to the tone in the "current" sample period.

With the sequence numbers to indicate lost frames, it is possible to drop three consecutive DTMF signaling subframes and not lose any information. If more subframes are lost, the receiver keeps on with the previous tones until new information arrives. A voice subframe received on the subchannel turns DTMF tones off.

After the last DTMF activity, the VFRAD sends only three more signaling packets. No bandwidth is taken for signaling during talk mode. This function resembles a normal phone, where pressing the tone keys breaks the voice path. You don't want to talk through tones, and voice may confuse the interpretation of tones. It is one at a time, not both.

This syntax contains no provision for supervision. The CAS transfer syntax is needed to signal on- and off-hook conditions. This omission should be corrected in the next version of the IA.

Dial Pulse or ABCD Bits (CAS)

The assumption behind this method of handling signaling is that any interface, analog or digital, is commonly found in digital format on T-1 and E-1 lines. Signaling there consists of up to 4 bits (ABCD) that reflect the state of the interface (up to 16 states). The bits are multiplexed into known positions associated with each voice channel in the digital signal.

For example, the A bit normally relates to the on- or off-hook status of the voice interface. The A bit may be the only one defined for certain older interfaces (like E&M). Under the first version of the VoFR IA, CAS transfer syntax is needed for supervision status even when dialing is DTMF.

Dial pulse information can be relayed by toggling this bit between 0 and 1 because pulse dialing is the same as pressing the hook switch very rapidly. Dial pulses are nominally 10 per second. A signaling bit repeats at 2000 per second (or more, at a multiple of 2000, up to 8000). So there are plenty of bits to describe 10 square waves per second.

Loop- and ground-start interfaces use two bits, A and B. The far end status is reported in more detail, for example "on hook, ringing."

In Chapter 1, the channel bank connected the analog loop to the digital transmission line. The CB converts the loop state to the 'channel associated signaling' bits. Digital interfaces necessarily have them already, so a VFRAD with a digital voice interface using CAS need only collect the signaling bits and package them into VoFR subframes at appropriate times. The subframe header payload type indicates a signaling bit transfer is in progress. There is only one additional subheader byte, that contains a 7-bit sequence number (bits 1-7) and an Alarm Indication Signal (bit 8).

The transfer syntax for ABCD bits is Annex B of the IA. It offers transport only. The IA places no interpretation on the bits–that's up to the VFRAD vendor. A vendor needn't invent much, fortunately–the specifications for channel banks include just what's needed.

When there is signaling activity (changes) the VFRAD notes the status of the analog interface every 2 ms and converts it to ABCD bits (however the vendor chooses). Signaling bits are accumulated for 20 ms, then sent in a subframe. Each subframe contains samples from three periods, a total of 60 ms, with a sequence number. The contents of successive subframes overlap, providing redundancy and protection against lost subframes similar to that provided by the DTMF syntax.

After there is no signaling activity on the channel (CID) for half a second, an update subframe is sent every 5 seconds, but without incrementing the sequence number, which remains "frozen" until signaling activity again occurs.

Common Channel Messages

This is the simplest. Transport is offered, within the PVC that carries the voice channels, as a pure data channel with a separate CID. The same rules about maximum message length and fragmentation apply to signaling as to any other data frame. See "Data Fragmentation" earlier in this chapter.

The most likely application is ISDN or proprietary message-oriented signaling between PBXs.

Other Needed Features

While not essential to the definition of VoFR, every practical implementation will need the following functions to achieve acceptable voice quality. Since these functions are performed locally, that is independently in each FRAD handling voice, generally they interoperate regardless of what the device at the other end of the connection does. Therefore these features are not included in the VoFR IA.

Echo Cancellation

A frame relay network is the equivalent of a 4-wire connection (separate paths for each direction) and so introduces no echo itself. However, the end points of frame relay VCs are usually 2-wire analog voice connections, like standard desk phones, for which one wire pair carries speech in both directions.

Any transition between a 4-wire path and a 2-wire path involves a 'hybrid' circuit, also known as a 'hybrid network.' It may be a special wire-wound transformer or the equivalent function in a semiconductor chip. Electrically a hybrid is a 4-legged bridge circuit–the 4th leg is often not drawn but in practice contains a variable resistor and capacitor to achieve balance against the resistance and capacitance of a specific 2-wire loop.

You will find one hybrid in the VoFR device and another in the phone at the end of the analog tail circuit (Fig. 26). Every analog phone has a hybrid

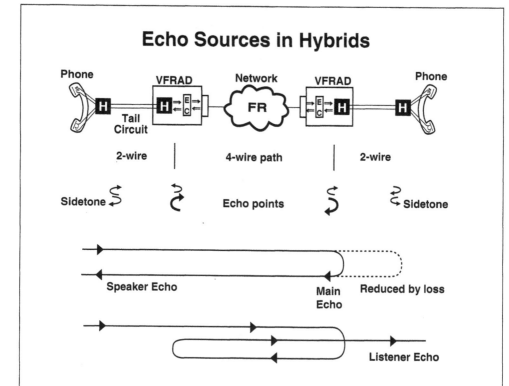

Echo Sources in Hybrids

Fig. 26. *The change or imbalance in impedance at the hybrid circuit, where a 2-wire connection meets a 4-wire connection, reflects sound energy back to its source. Callers hear the reflections as echoes. Most of the echo occurs in the VFRADs. There the frame relay transmission link (digital, 4-wire equivalent) meets the 2-wire analog tail circuit.*

to combine the four wires from the microphone and earpiece (2 each) into the 2-wire local loop. The difference in impedance (combination of resistance, inductance, and capacitance) at each side of a hybrid causes a reflection. Part of the voice energy entering a hybrid on the 4-wire side is sent back to the source (on the 4-wire side), creating an echo. Hybrids inside phones are not adjustable.

With as many as four hybrids in a voice circuit, there can be multiple copies of a speaker's voice bouncing back to him.

The largest echoes are on the ends of the 4-wire circuit portions of the path: from the 4-wire back to the 4-wire (large loop-back arrows). These are the two critical echoes to cancel. Fortunately, both pass through a voice processing DSP in a VFRAD, where they can be cancelled digitally. The same DSP that compresses the voice can also cancel echo.

An echo canceller (EC, Fig. 27) learns how the attached tail circuit (which includes the VFRAD's own hybrid) reflects the signal received from the FR network. The EC watches it and adjusts the adaptive filter to mimic the reflection. This is done in less than half a second at the beginning of a connection. The adjustment includes finding both the delay and the size of the reflection from all points on the tail circuit, starting with the hybrid in the VFRAD and including gauge changes in the loop, bridged taps, the phone, etc.

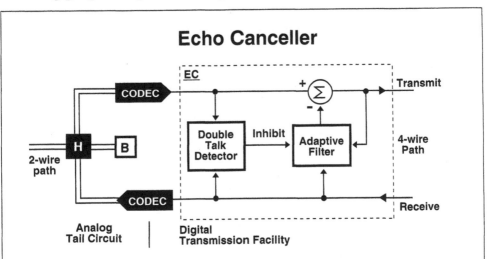

Echo Canceller

Fig. 27. The EC learns the characteristics of the tail circuit and how echoes return: amplitude vs. frequency vs. delay time. The EC saves a copy of the transmitted signal, applies the response it has learned, then subtracts the resultant portion of the delayed "receive" signal from the "transmit" signal to cancel the echo of the "received" original as it returns from the tail circuit.

Knowing that information, the EC can use any new incoming signal to generate the appropriate cancellation signal. The adaptive filter adjusts the strength of the received signal, delays it, inverts it, and sends it to the summation circuit (adder), which subtracts it from the tail circuit transmission toward the FR network. After the filter 'converges,' the signal fed to the summation circuit just cancels the echo. Otherwise that echo (or echoes) would be heard at the other side of the FR network.

Echoes back onto the 2-wire loops, from the phones, are less of a problem. These echoes are reduced because analog paths have natural signal losses that reduce the echo's loudness. Additional 'return loss' may be inserted deliberately to mask the echo.

Immediate echo heard by the originating speaker is not harmful. In fact, some echo is designed into a telephone's hybrid so the talker can hear himself–this is called 'sidetone.' When the analog tail circuit is short (within the same building), echo from a hybrid in the VFRAD on the near side of the network tends to be heard as sidetone also.

The most disturbing echo occurs when a speakers hears his own voice signal after it bounces back from the far side of the network. Echo becomes a problem if the delay exceeds about 1/10 second–which is true for almost every case of VoFR. When echo is loud enough, and delayed a large fraction of a second, it can prevent some people from talking at all.

The high quality of digital transmission actually can add to the echo problem: the frame relay network has no signal loss that would reduce echo loudness. At the same time, frame switching inherently adds latency so the transit delay will be increased (compared to an analog line).

Total delay across a frame relay network consists of compression processing (10-50 ms), switching time in each frame relay node (5-10 ms each), and actual cable length divided by the speed of light. In practice, electrical propagation speed is about 1 ms per hundred miles. It is slower in copper wire or glass fiber than in vacuum.

To preserve voice quality, a VoFR voice connection needs echo cancellation at both ends. The EC at one end cancels talker echo that would be heard at the other end and helps control listener echo at its own end. Note that the delay jitter in the frame relay network, if any, isn't noticed by the echo canceller–the EC works on echoes from the fixed-delay tail circuit.

How much incoming signal the canceller can remember determines how long the tail circuit delay may be and still achieve cancellation. The limit is imposed by available memory in the adaptive filter to hold a copy of the received signal. In most cases a window of 10-20 ms is sufficient. That time represents the round trip delay over tail circuits of 500 to 1000 miles. However, 20 ms may not be enough to cover listener echo.

For VoFR applications the long distances should be on the FR network; analog tail circuits should be minimized, and ideally kept within a building. Longer tail circuits probably indicate a need for another frame relay access point, which would probably be cheaper, if available, than the long analog tail circuit.

Silence Suppression

To reduce the traffic on the FR network, a VoFR device applies voice activated transmission (VOX), also called silence suppression or 'voice activity detection.' That is, during a sample period (one 10 or 20 ms interval, for example) when the sound input never rises above a threshold, no voice frame is sent to the backbone network. During silent periods, only HDLC flags are sent by the VoFR portion of the access device.

A comparable process for data is 'spoofing' of polled protocols by a FRAD. This feature prevents sending frames across the network unless they contain user information; polls are stopped by the FRAD at the host and generated by the FRADs at remote sites.

Loud background noise (or music on hold) can fool the machine into sending additional, unnecessary frames, reducing the system capacity. Smarter VoFR servers set thresholds dynamically to minimize the impact of background noise. Really smart ones might know the difference between voice and music.

There is a Payload Type called 'silence indication' that may be set by the VFRAD in the subframe header if the transfer syntax for that connection makes use of it. An appropriate use for the silence indication would be in the last subframe header that contains the last portion of audio at the end of an utterance. The receiver then knows it must deal with a temporary halt in the subframe stream on that channel. Frames marked as silent also tell the receiver what the background noise is at that time.

If a voice call connection is still established, the receiver should insert noise during periods when no subframes are received. Audible noise prevents the listener from thinking the line has gone dead or the connection is broken.

Noise may be based on a sample of background detected during the conversation. For example, the frame with PT = "silence indication" could be repeated until voice frames resumed. Or the VFRAD might generate white or pink noise (generic static-like noise). The VoFR transfer syntax for PCM encoding lets one FRAD tell the other the exact loudness of the background noise, to prevent 'pumping' or a shift in background level between talking and silent periods.

The statistics of speech regarding pauses in most conversations make silence suppression very appealing. The VOX feature is the same as Digital Speech Interpolation (DSI) and produces the same decrease in average bandwidth per channel. Voice activation can double the capacity of a line at no additional cost in voice hardware and little in software.

With VOX, the average bit rate for an 8 kbit/s CELP connection drops to less than 5 kbit/s. At that rate, a 56 kbit/s access loop (to a frame relay network) can carry 4 to 6 conversations and still allow for a few interactive data terminal applications. With LAN connections or multiple fax machines at a branch office, it might be advisable to have no more than four, or even two, voice channels on a 56/64 kbit/s line.

VOX benefits central sites, too. VOX can halve the T-1 line count from the frame relay network for a given number of active voice channels. Again, there is a greater potential to share those access links with data applications when VOX is applied to reduce the need for voice bandwidth.

Processed digital information (compressed voice) is stored in memory, at least briefly, for every interval, including silent ones. When an utterance starts (that is, when the volume exceeds the threshold set for VOX) the sending station can reach back in time (dip into memory) to include the start of the speech segment. Thus there need be no 'clipping' at the beginning of an utterance after a period in which the speaker is silent.

Reaching back in time will delay the relative playback time at the receiver by the period picked up from memory. Fortunately the information is in memory, so there will be no additional accumulation delay. Since this happens only at the start of an utterance after a silence, the listener has no way to detect the additional delay.

Delay Compensation

In the network, every frame switch takes time to process a frame. In the frame relay backbone, a switch should delay a frame about 5 to 10 ms. For comparison, an older X.25 switch may add as much as 60 ms to the 1-way trip time. The other major component of delay is signal propagation (1 ms per 100 miles). CPE processing time, for example compression and decompression, is additional.

Experience so far indicates that transit delay for public frame relay networks doesn't usually exceed about 100 ms. That is low enough to keep conversations in "nearly real" time. However it is long enough to require appropriate echo cancellation.

Inherent in compression is the fact that it takes less time to transmit a voice frame than the period of speech represented by that frame. Recall from earlier in this chapter the calculation of transmission time for one subframe containing 10 ms of G.729 voice: about 2.3 ms on a 56 kbit/s port. If there is only one voice call active, there will be one frame every 10 ms. So each frame is separated from the next by 7.7 ms, during which time the line is idle.

If all these delays were constant, then voice frames created and sent at uniform time intervals would arrive at the same 10 ms intervals (or 50 ms intervals if using a packing ratio of 5). We would again have an emulation of a current loop (similar to what ATM attempts with 'circuit emulation service'). However, there will be some variability in delay (jitter) during congestion as frames in busy switches queue up to get on an outbound transmission line. Introduction of long data frames on the same transmission paths will also increase jitter.

Delay is a particular concern where VoVR networks interconnect with a public network. In some markets the regulatory authority (typically the PTT or the dominant carrier) imposes voice quality standards that penalize or limit the allowable delay.

Processing Delay

Processing delay is a significant measure of a compression algorithm. Processing delay is the time the DSP takes to compress the linear PCM digitized voice segment (20 ms worth, for example). The time is measured between the end of a sample period and the time when the compressed frame is ready to send. Delay depends on both the complexity of the algorithm and the speed of the DSP. Typical values range from a high of 40-70 ms down to 5 ms.

Processing delay is added to transit delay to reach overall delay--which is what counts in perceived voice quality. If there is too much delay between speakers, both may start talking at the same time. They won't realize for some syllables that they are both talking. These "voice collisions" are common on satellite circuits where total round-trip delay is at least 1/2 second. They are not a problem if total delay is under 100 to 150 ms.

One of the attractions of G.729 CS-ACELP is that the delay is moderate. Other algorithms are designed specifically to minimize delay: G.728 Low Delay CELP, for example, has a 5 ms sampling period.

Delay Jitter

Congestion in the frame relay network increases transit delay as a frame waits in a buffer to access the outbound transmission line. The effect of random congestion delay is to lengthen and shorten the intervals between successive voice frames. To date, the loading on most public frame relay networks has been light, so queuing jitter has been small. As networks fill up with additional subscribers, most analysts expect rising congestion to increase jitter.

A delayed frame means it will arrive later than expected. Voice frames following the first delayed frame may back up behind the first at any queuing point in the network where the added delay exceeds the interval between voice frames. When finally forwarded, all frames in the queue will be sent contiguously, with only a single HDLC flag separating one frame from the next.

In a sense, the second and later frames in the "clump" arrive earlier than expected, relative to the first frame, because they are delayed less than the first frame. In other words, the time interval between two consecutive voice frames when sent may not be the same as the interval between them when received.

The receiver accepts frames from the network as presented in a clump and puts them in a buffer. If jitter in the network exceeded 20 ms, then two or even three frames might arrive in about 10 ms, but they would take 20 or 30 ms for the receiver to play them back. With packing, the interval between frames increases by the packing factor, making them less sensitive to small jitter.

To play back frames, the receiver takes them from the buffer, one at a time in the order received, decompresses the information, then converts from a digital to an analog signal.

Compensation for variations in transit time through the FR network (jitter compensation) depends largely on an initial delay introduced by the receiver. The strategy is relatively simple: the receiver waits for a small, fixed time after the first frame arrives, but only if the receive buffer is empty at that arrival, and no voice frame on that particular connection is being played back. That is, the receiver is silent when the frame arrives.

This waiting period is set to the expected value for variability in transit time across the network, and may need to be adjusted occasionally. For example, if the transit time across the network is 75 ms +/- 15 ms, then the expected variability is 30 ms.

With such an initial wait, the receiver almost always should see another frame arrive before the first is finished playing. A second or subsequent frame is played out immediately after the frame before, with no wait added, if the receiver sees a frame with the same channel ID in the buffer before finishing playback of the previous frame.

In this way the listener hears a continuous reproduction of the speaker's voice. Any variation in delay caused by congestion, up to the fixed delay, is hidden from the receiver.

An assumption is made that the compression algorithm decompresses and plays back each received frame as a separate unit, with playback time determined by the length and content of the frame, not the time it was sent or received. For contrast, consider PCM encoded voice where each byte received on a TDM channel is converted immediately to an analog value.

For a visual metaphor, imagine a voice message being taped by the speaker. As the sending end, the tape is cut periodically (every 50 ms) and the segment is sent by mail (delay is variable). At the receiver, the first delivery of a tape segment tells the listener to warm up the tape player. After a short time he feeds in the first tape segment. The expectation is that the second tape segment will arrive in the mail before the first segment finishes passing into the tape player. If the next piece of tape arrives as expected, the two segments are spliced together and they play out continuously. Subsequent segments are spliced on immediately (put into the FIFO buffer). Even if several tape segments arrive at once, and are spliced immediately, the playback will still sound normal as the reconstructed tape feeds into the player.

If the receiver runs out of tape, he could just stop the playback machine and wait for another "first segment" to arrive. Smarter algorithms watch for a "silence indication." If the last frame in the buffer doesn't indicate "silence follows," the playback VFRAD tries to fill in for a while, expecting another frame shortly. It might "hum" for a short while (because it doesn't know the words), extending the last sound received. Or it could fill in with background noise.

The Fax Transfer Syntax (subframe format) and some voice transfer syntaxes will include a sequence number in the subframe header. The fax sender increments the SN at fixed intervals, making it a relative time stamp as well as a check mechanism for lost subframes. Most voice algorithms, including G.729 and others suited to VoFR, don't require a sequence number.

Timing Between Speaker and Listener

Note that the queue handling designed to compensate for jitter, described above, does not preserve the absolute time relationship from when the speaker utters a word to when the listener hears it. Rather, the timing that is preserved is "local" (in time), relative to other words in the same 'utterance' (spoken at the same time, between two adjacent silent periods).

1. The VoFR process described can introduce pauses, or lengthen silent periods when VOX is active. If the first part of a stream of voice frames (from the same utterance) crosses the network quickly, and congestion creates a delay for the middle frame, then a pause or silence may be audible to the receiver. The VFRAD may create sound to bridge the time gap.

2. The process can eliminate pauses of a certain length because frames in the buffer are played out without pause between them. Thus a "silent" period, up to the length of the delay variability through the network, may disappear on playback.

The first case produces a perceived silence at least as long as the initial delay, some tens of milliseconds at most. When the receiver starts playing again, the receive buffer will be backed up with successive frames so playback will not be interrupted again during this utterance–unless the congestion delay exceeds the previous delay by more than the initial delay. That is highly unlikely in a frame relay network that is providing satisfactory service in the first place. Result: a small possibility of one small "silence insertion" per utterance.

For the second case, a VOX parameter at the sender can be configured to send some "silence" before halting the transmission of frames. For example, if 100 ms or ten 10 ms subframes of "silence" were sent after each utterance, that time would exceed the variability in delay across most FR nets. Such a setting would ensure that short pauses are preserved exactly. Result: long pauses may be shortened by some tens of milliseconds, which is undetectable in normal conversation.

This process ensures that the listener hears playback sounds correctly, even though some segments (parts of utterances) may be shifted in time relative to others. To the listener, there is no way to tell the difference between this time-shifting and a speaker who varies the length of pauses in speech. During the next silent period, the receiver's buffer empties and "relative time" returns to normal.

By holding many frames in queue, and playing them out at a rate determined by their contents, the receiver avoids the need for:

-- time stamps on each frame,

-- complex algorithms to adjust for network delay, or

-- expensive modifications to existing networks in order to give voice frames priority over data frames (priority must be assigned by the originating VFRAD).

The two worst cases, a very long pause or lost frames, will merely cause the listener to ask the speaker to repeat something, also known as the "Huh?" protocol.

Features Still Proprietary

To speed publication of the VoFR IA, the FRF-TC decided to present the IA in stages. The first release does not include, for example, switched virtual circuit procedures that had been proposed earlier. They should return in the second edition.

Several vendors offer the following additional features which are not covered by the IA but may be of interest to network operators.

CIR Enforcement

In a frame relay service, the backbone network recognizes only the main FR address. The subframe structure of VoFR is hidden from the switches. By definition, any bandwidth parameter can be assigned only to the DLCI or PVC as a whole. All of the channels based on subframes, up to 255 of them, must share one allocation of 'Committed Information Rate' (CIR). If the network ever has to enforce a limitation to the CIR, it will treat all frames with the same DLCI alike.

It is up to the customer's equipment, the VFRAD, to enforce the bandwidth constraints when and where the frames originate. After that, it's too late to distinguish between voice/fax and data channels with the same DLCI.

The frame relay service definition at the User-Network Interface (UNI) has its own rules:

– the network commits to carry a certain amount of information per second (the CIR) to its destination regardless of what other network users do.

– The CIR may be less than the speed of the access line. Since a frame must be sent at the line speed, the network commits to accept a fixed number of bytes (the committed burst, Bc) in a continuous block and treat them as being within the CIR.

– A larger 'excess burst' (Be) size may be configured on the UNI. 'Be' is set larger than 'Bc' and equal to or less than the line speed times the measuring interval (usually 1 s). Information sent above 'Be' may be discarded immediately by the first switch as a way to enforce traffic management. Other carriers set 'Be' to the line speed but say 'Be data' are carried only if capacity is available.

– Frames carrying bytes above the CIR or beyond the committed burst, while still below the 'excess burst' (Be) level, will be marked by the network as 'discard eligible': a switch changes the DE bit in the header from 0 to 1.

– When the network is congested, it will discard first those frames marked with DE = 1.

Changing the DE bit is a minor task. Each switch, in general, changes the address of every frame that passes through it. Any change in the header or payload requires a recalculation of the CRC, but that is done by the transmission hardware automatically for every frame.

Network designers often recommend that any frame relay virtual circuit that carries voice be assigned (that is, you pay for) a Committed Information Rate (CIR) at least as large as the sum of the encoding rates for all voice channels on the VC. Four CELP channels encoded at 8 K each should have a CIR of 32 K at the UNI, more if there are priority data channels too. As the number of channels increases (certainly by 24 voice connections on one

PVC) you can rely on voice statistics and silence suppression to reduce the minimum CIR. For 24 channels you might try a CIR of 128 K, and adjust it based on experience.

Why?

If all voice channels are active, a load of 32 kbit/s, the VFRAD will be able to send all voice frames immediately and not exceed the CIR. If those voice channels are the only traffic, none of these frames will be DE. The network's commitment then guarantees (almost) that the voice frames will be delivered.

But what happens to data when voice grabs all the CIR? There are several options:

– Halt all data transmission temporarily, until a talker pauses or hangs up.

– Segregate voice and data into separate frames (no voice subframes packed with data subframes in any "mixed content" FR frames). If the VFRAD marks data frames as DE, the network does not count them against CIR usage. That way voice frames remain within 'Bc' while data frames get to use 'Be.'

– Buy more CIR, more than all the voice channels could use. Then there will always be some Bc capacity for data.

– Put voice and data on different DLCIs. If voice and data don't share a DLCI (PVC) on a VFRAD, each traffic type may have its own CIR.

Most network operators have opted to minimize carrier charges for PVCs by insisting that a small remote site be assigned only one DLCI. Monthly charges for additional DLCIs range from $1 up to $50. Financial discouragements to the use of more DLCIs are sometimes tied to limits in the carrier's switches. Tariffs and equipment change, however, and you can negotiate.

VoFR subframes are the first industry-wide response to the "one DLCI" networking requirement. Previously we had no standard method to multiplex voice and all kinds of data over one PVC. The Multiprotocol Encapsulation implementation agreement, while close, excluded X.25 traffic, which has its own specificaiton for encapsulation in frame relay. Hardware vendors put together proprietary solutions that did not interoperate. VoFR truly handles any data protocol.

Prioritization

To ensure fair sharing of the bandwidth available for network access, the VFRAD sets up and enforces a set of rules. The consensus among VoFR hardware vendors is: voice first, voice next; then data. Fax is treated like voice when it uses a "voice" channel ID.

Voice/fax frames in today's equipment are always transmitted to the network before data frames. If there is a voice frame ready to send, it goes next, immediately after whatever frame may be in the process of transmission. Data frames are not aborted to give a voice frame access, but data wait regardless of how many frames are ready or how long they have waited.

How this is done is neither obvious nor trivial. In witness, all VFRAD vendors seem to have slightly different rules. Recent testing by customers evaluating FRADs and routers shows a very broad range of capabilities in setting priorities, from dismal to excellent. Any flexibility available to the user will likely be needed at some time.

Such flexibility might include a temporary reduction in priority for a connection when it has used up its allocation of the CIR. That change will give another connection higher priority temporarily, but not block continuing transmission if there is no other traffic. There is also a need to recognize all conditions and priorities on a per-DLCI basis.

It is up to the VFRAD vendor to have a strategy for priorities among data types. For example, should SNA traffic have higher priority than TCP/IP traffic? Main frame users often demand such priorities, saying delayed frames cause a terminal session to time out and be disconnected. Re-connection usually isn't automatic.

The time-out argument is dealt with by 'spoofing' in the VFRAD for SDLC and other legacy polled protocols. Instead of letting the host computer send polls to all the terminal controllers as in the past (on analog multidrop lines), the VFRAD stops the poll and issues an acknowledgement to the host. The VFRAD at the remote site generates polls that look like they come from the host computer. Both the terminal controller and the host remain happy, no matter how long it takes a data frame to cross the network. Of course there are arguments based on productivity or response time that may justify prioritization for certain connections.

Fragmentation of data frames (discussed earlier in this chapter) plays a key role in prioritization. Usually there will be many short intervals when no voice frame is ready to send. These are opportunities to insert short data frames or fragments of longer frames without hindering voice traffic.

Congestion Management

In the event of congestion indications from the frame relay network, the VoFR equipment, like any well-behaved FR terminal, should immediately reduce the amount of traffic sent on the affected DLCs. If the FRAD had been using excess burst (Be), that should stop. No more than the CIR and committed burst (Bc) should be sent. If the BECN signal continues for more than a few seconds, the terminal should drop back again, to even less than the CIR.

This may mean that data is squeezed if voice channels are active, and voice may have a problem too. Vendors wrestling with response strategies may reduce voice encoding rates (CELP) or drop some bits from frames (E-ADPCM). Others may simply ignore the congestion, continue to send full tilt, and let the network choose what to discard.

Most implementations of prioritization are not sophisticated enough to deal with such a complex situation involving many subchannels on a PVC. They choose the last option and don't reduce transmissions. But then they lose control over prioritization–the network picks which frames to discard, almost randomly.

The problem arises because the network sees only the frame relay address (DLCI) and does not understand which of the subchannels is causing the congestion or understand their relative importance. The network cannot enforce flow control selectively by CID.

It is up to the VFRAD to know the sources of traffic and how to apply flow control to each. When a VFRAD puts many voice channels and several different types of data into a single PVC at a branch office, the problem is quite complex. There are no such implementations based on standards at this time, all are proprietary.

The most thorough prioritization scheme (FastComm at this writing) can enforce the CIR limit, prioritize by protocol type, protocol address, or physical port, and prioritize frame relay PVCs. When necessary to reduce traffic sent to the network, the "working" CIR is reduced temporarily while preserving all relative priorities. As an alternative, or when ongoing congestion indication forces a further cutback, the strategy may change. Certain classes of traffic, like voice and SNA data, may be given almost a guaranteed bandwidth regardless of congestion indications. Other traffic will be severely slowed but not blocked entirely.

Watch for new implementation agreements, RFCs, and standards as the industry gains experience with crowded networks and various ways to prioritize traffic and manage congestion.

Chapter 4

Voice Signaling

Transmitting voice in digital form along a fixed path is relatively simple. That's what a channel bank has done for 30 years. There are standards like the G.7xx series from ITU that define the way voice is digitized and compressed. These methods or algorithms are available from many vendors as software code or hardware modules. But voice compression and transmission are not all you need for voice over frame relay.

Setting up a path on the frame relay side, finding the route for that path, indicating when to ring a phone, and interpreting dialed digits is another story: signaling. It is a necessary but different form of information transfer. This is the proper domain of the PBX or central office voice switch (the class 5 switches like the Lucent 5ESS or Nortel DMS-100).

On a global scale, the complexity is immense. The corresponding protocols to handle the variety of information, for use within telephone companies, are generally standardized by the ITU, like Signaling System 7 (SS7). But signaling schemes between the customer and the carrier tend to be national or regional.

To reduce complexity, vendors of voice over frame relay have treated signaling as a matter for a private network, not the global PSTN. The result is proprietary formats, often with functionality limited to simple pass-through of the interaction between a telephone and a PBX.

The more sophisticated VoFR systems will interpret dialed digits and map a call request to a frame relay PVC that exists before the call is placed. At least one vendor bases a call setup on switched virtual circuits, but only within the backbone of switches from that vendor alone. SVCs are not seen to date by the user of voice equipment (phones and PBXs). When SVCs are available from carriers, the range of applications will increase dramatically, but signaling itself should become much simpler–the route-finding function will be taken over by the FR network. The calling phone equipment will not have to know or learn a path. It will simply give the called number to the network and expect a connection to the right place.

Usually signaling bandwidth isn't included in the nominal bit rate for the voice encoding. Thus an 8 kbit/s voice channel might need another 1 kbit/s for signaling. Just as often, signaling and voice are not active at the same time, so the bandwidth allowed for the voice path is more than enough for signaling messages that precede and follow a conversation. Fortunately, frame relay allocates bandwidth on demand, elastically, so any differences between voice and signaling bandwidth are accommodated easily.

While almost all VoFR equipment offers analog voice ports, with the functions described below, the signaling must be converted to digital form and sent as VoFR subframes over the backbone network. Each vendor has a proprietary way, but all are expected to migrate to the FRF IA (Appendix C)–at least for the simplest connections.

Channel banks give each state of an interface a distinctive code, based on the ABCD bits in T-1/E-1 signaling. Examination of the standards documents reveals that the AB bit values are different for different types of signaling. That is, 'on hook' for FXS is indicated by AB = 11, while for the E&M interface the on hook value is 00.

However, there is an exception for 'special access' which inverts the signaling bits to make FXS similar to E&M. If a unified designation for 'on hook' is ever adopted, 00 seems justified.

Some proprietary features may continue to require proprietary signaling into the foreseeable future.

Analog Voice Interfaces

Signaling started when all we had was the analog loop. One circuit conveyed all the information there could be:

– starting and ending a call, on- or off-hook status, by starting and stopping a direct current,

– arrival of a new call, ringing, by imposing a high-voltage alternating current,

– rotary dialing by interrupting the d.c. current, and

– tone dialing by transmitting tones the same as voice.

Certain analog voice interfaces, the E&M versions, are more complex. They have additional "handshake" circuits for signaling and require the ability to convey the status of multiple wires, often with critical timing relationships.

Despite the almost total internal conversion of the PSTN to digital switching and transmission, the great majority of lines remain analog. Most of the voice equipment installed today in North America has analog interfaces, either on the trunk (CO) side or the line (telephone) side. Europe has adopted digital interfaces almost universally, in the form of ISDN, for practically all new installations of any size. In the US, a company may grow to 30 or more people before justifying a digital PBX.

Branch offices, particularly, often need no more than two to six voice connections. At that size, the need is met most economically with an analog key system rather than a digital PBX. Thus the greatest number of sites using VoFR will need analog ports on the VoFR FRAD (VFRAD). What follows is a review of voice grade interfaces likely to be encountered in North America. There are dozens of other forms of signaling at analog interfaces in use worldwide.

[Readers are invited to nominate additional interfaces for the next edition.]

In each section, the possible states on the analog side are also translated into signaling bit patterns. It is these bits, created from the states of the analog interfaces, that become the robbed bits (ABCD bits) tucked into the least significant bit position of every 6th sample of PCM encoded voice in a T-1 extended superframe. They are similar to the channel associated signaling (CAS) bits multiplexed into time slot 16 of an E-1 circuit. Voice over Frame Relay collects these ABCD bits into separate signaling frames.

Originally there were only the A and B signaling bits, in the 12-frame T-1 Superframe. There are four positions for signaling bits in each 24-frame Extended Superframe (ESF). When an interface generates only AB, those values are repeated in positions CD; C=A, D=B.

BORSHT and Other Definitions

In describing analog signaling, certain terms are needed. Classic functions of analog interfaces on CO lines have long been abbreviated as BORSHT: Battery feed, Over-voltage protection, Ringing, Supervision (now often referred to as Signaling), Hybrid function, and Test access.

Battery: From the early days of telephony, the phone company supplied the power to operate phones. The current source has been a nominal 48 volt d.c. battery (24 lead/acid cells, 2 V each).

The term "battery" is still descriptive: there are large banks of lead-acid wet cells in most central offices. That's why the phones still work when your local power fails. Unless otherwise specified, the positive (+) battery terminal is connected to ground (earth). This makes 'battery' a negative (-48 V) signal with respect to ground.

The full-strength interface defined for channel banks (in a 1985 document from AT&T, Pub. 43801) assumes they will have to power a phone through a cable more than 3 miles long. Voltage may be less (as little as 15

V) from equipment like a VFRAD if it is designed for short cable lengths. This will apply within a building.

Overvoltage protection: The components in the phone like the microphone can be destroyed by high *voltage.* Wires on poles act as windings on a transformer or receiving antennae and react to nearby lightning, even if not struck directly. Fortunately the *power* and *energy* are usually low enough to be absorbed by a small component. The first ones were blocks of carbon; modern devices may be gas discharge tubes or semiconductor diodes. Minimal protection is needed if the analog tail circuit is within a building.

Ringing: In North America, the central office switch supplies a 20 Hz AC signal on the local loop toward a telephone, 2 s on and 4 s off, to alert the subscriber that a call has arrived. Voltage may exceed 100 V rms (on top of 48 V d.c.), with a minimum of 40 V rms (sine wave). In other geographical areas the voltage, frequency, and cadence may vary. British movies depict their phones with a double ring cadence that reduces power consumption.

FRADs may be designed for attachment to customer premises equipment only. If so, the ringing voltage supplied would be near the minimum. Within a building there is far less voltage drop in the wiring than if the FRAD were in the central office and had to ring a phone over 18,000 feet of cable.

Supervision: When human operators still said "Number please" (before dial tone) supervision consisted of a lamp on the switchboard that lit when a phone went "off hook." The operator plugged into the jack next to the lamp to speak with the subscriber making the call.

Picking up the phone (going off-hook) causes the hook switch inside the phone to make an electrical contact (turning on the loop current). Refer to the drawing in Chapter 1. The hook switch was named when there actually was a hook on the side of the wall-mount phone on which the handset was hung when not in use. Don't be confused: "off-hook" means the loop current is "on."

When the hook switch closes, it completes a d.c. path inside the phone, which draws loop current from the battery. That current is the one which lit the lamp to alert the switchboard operator a subscriber wanted to place a call. Now the same current triggers an electronic response.

Rotary dialing was able to control call setup by interrupting the loop current. Each break or pulse in the current caused an electromagnet to move a stepping switch to its next position. Today, electronic circuits count the pulses or, increasingly, interpret the dialing tones from pushbutton phones. Supervision or signaling is essentially the same function, except that supervision may refer more explicitly to reporting the on/off-hook state of a phone.

Hybrid: The typical local loop is only two wires. Yet many parts of the phone network need separate electrical paths in each direction: four wires.

A phone contains separate microphone and receiver, each of which have two wires. Yet both components must connect to the loop. Long distance transmission must be amplified, but amplifiers that are unidirectional. Consequently there must be a separate 2-wire path in each direction (making it a 4-wire path). The solution is called a hybrid, an electrical component that couples a 2-wire loop to something with two paths on four wires. Hybrids used to be special transformers, but new ones are integrated circuits.

Test access: Analog lines can develop problems that are most easily isolated if the test equipment can be attached directly to the copper wire. This access might be a connector on the terminal block where the line enters a building, rather than part of a phone. Central office equipment almost always has jacks where testers can plug into the copper loops. Digital transmission is not bothered as much by the impairments that worried analog engineers. Most digital circuits test lines and make adjustments automatically, minimizing the need for access. Direct metallic access is now seen outside the CO mostly on high-end CSUs that terminate local loops.

Talking specifically about the signaling procedures requires several more terms:

Ground: may be earth or equipment chassis ground, but also may be a separate electrical lead (wire strand) in a cable or at an interface grounded at the far end. It is for signaling only (not for power transfer).

Battery: -48 V with respect to either earth or signal ground.

Open: switch is open, lead is not attached to ground or battery.

Closed: switch on balanced pair is closed.

Asserted: whatever condition indicates off-hook.

Idle: whatever condition indicates on-hook; normal.

Line: from a switch, a 2-wire path defined to an extension telephone.

Trunk: 2- or 4-wire transmission path, possibly with separate signaling path(s) as in E&M, between switches.

E&M Signaling

One of the most common interfaces for PBXs and switches is E&M. In addition to the voice path, there are two signaling circuits, one to send and one to receive. Depending on your source, the abbreviation stands for:

– Earth and Magneto, early terms for ground and battery.

– E(ar) and M(outh), indicating what the switch or PBX equipment does on each lead when "talking" on the multi-wire interface to the FRAD or transmission line (Fig. 28);

– connector block positions that were marked alphabetically, E and M, rather than numbered, used years ago to install this type interface. The positions were codified in a standard 'practice' or instruction set for telephone company craftsmen, thus preserving the name.

Note that E&M leads may consist of either 2 wires (plus ground lead) or 4 wires (two balanced circuits). However these are not the wires referred to when calling an E&M interface '2-wire' or '4-wire':

E&M interfaces are characterized as "2-wire" or "4-wire" by the nature of the voice path, not the signaling leads. That is, voice can be carried on either a single loop of one pair of wires or on a separate wire pair transmitting in each direction. E&M modules are available to support both 2-wire and 4-wire voice paths.

As shown below, there are defined five different types of E&M signaling circuits, written as Roman numerals I to V. The number of physical wires used for signaling depends on the "Type."

As noted for Type IV E&M, below, two VFRADs may be hooked together directly (back to back) if the E lead on one VFRAD is connected to the M lead on the other (the voice path must be connected as well).

Types I and III need an auxiliary circuit between them, to provide battery as well as cross-connect the E's to M's. Types II, IV, and V are easily connected together if the sB, M, sG, and E of one side are wired to the sG, E, sB, and M leads of the other, in that order. Ignore the sG and sB on Type V.

Off the shelf interface converters are available to connect an E&M with practically any other interface, including another E&M. Some types and mixed types may have to be "kluged" together by adding a battery, ground, relay, etc.

E&M Interface, Switch To Trunk

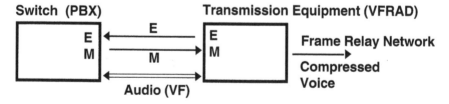

Fig. 28. "Ear" and "Mouth" refer to the switch when it talks to a transmission facility or trunk line. The voice path may be two or four wires; there are five arrangements for the E&M circuits. Each signaling lead has the same name at both ends, whether that connection sends or receives the signal.

The "signal" on the E or M lead is any of several conditions, again depending on the Type of E&M interface. The conditions called ground, battery, open, and closed are defined earlier in this chapter. Typically, only two of the conditions are possible on any specific lead, one for each possible state of that lead: idle or asserted.

Note that signaling and hook switches in the diagrams that follow are marked N.O. (normally open) or N.C. (normally closed). In this section, "normal" means on-hook or idle, the state in which phones spend most of their lives. When a device goes off hook, the N.O. contacts close; the N.C. contacts open.

The voice Frame Relay Access Device (VFRAD) is always assumed in this section on E&M to be transmission equipment, like the channel bank, and not the voice switch. It is the switch that has a mouth and ear. Therefore the E lead carries a signal from the VFRAD to the PBX, and the M lead allows the PBX to signal the VFRAD.

In cases where there is another E&M interface at the far end of the transmission line, any E&M signaling system conveys the M signal on one end to the E lead on the other. In this way there is a full duplex signaling path between the two switches or PBXs.

In the traditional telco world of channel banks and T-1 lines, the two states of the M lead (idle or asserted) are the only signaling messages. They translate into the two signaling states of the A bit: 0 = idle, 1 = asserted.

When the E&M interface is carried on frame relay, the FRADs on each end that are part of the signaling path have the same lead designation (E or M) as the attached PBX:

(E on PBX) <E on VFRAD) <frame relay net] <M on VFRAD) <M on PBX)
(M on PBX> (M on VFRAD> [frame relay net> (E on VFRAD> (E on PBX)

E&M is used in local loops. But rather than run so many wires (up to 8 for one voice circuit) over any great distance, the phone company uses signaling conversion devices to convey the state of the M lead on each end to the E lead at the other end. A VFRAD performs this function on a frame relay network.

On analog circuits, the state of the M lead is converted to a tone signal carried on the voice path. The tone doesn't interfere with callers because the presence of the tone indicates the line is idle: the tone must be absent if the circuit is in use ('seized' or 'off hook').

Often a type of E&M interface at one side is allowed to interact with some other interface, like loop start. These applications may use proprietary features.

The E&M interface is defined from the switch to transmission equipment, with the assumption that there is another switch at the other side of the transmission trunk. For the PBX to seize the trunk, it asserts the M signal when the E lead is not asserted (E asserted means the trunk is busy, or seized from the other end). When the far switch responds, by asserting its M lead, the originating switch sees that response signal on its E lead.

If dialing information is to be transmitted between the switches, the called end may indicate its readiness to receive digits in one of two ways:

— **'Wink start'** ROM: The called end, when ready to receive dialing information, sends back a 'wink' (Fig. 29). This is a signal of at least 140 milliseconds, asserted on the far-end M lead, that appears on the originating end's E lead. The Called Switch also may return audible dial tone on the voice path at the same time. The ANSI standard for PBXs (EIA-464)

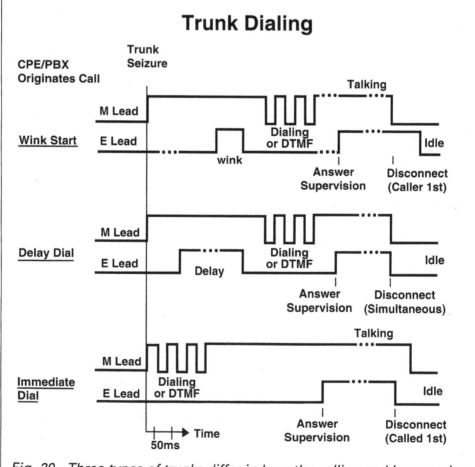

Fig. 29. Three types of trunks differ in how the calling end knows when to start dialing.

allows dial tone on the voice path to be substituted for the wink signal, if the other PBX will accept it; a configuration issue. The calling end then dials, either by toggling the M lead or by transmitting DTMF tones on the voice path.

— **'Delay dial'** trunk: when the called end sees the E lead active, it immediately (<300 ms) asserts its M lead (goes 'off hook') until ready to receive dialed digits. When ready, it returns to 'on hook' and may or may not supply dial tone. The calling end then dials, using pulse or tone.

When the called switch is not busy and can respond quickly, both wink-start and delay-dial will behave similarly and may be difficult to tell apart. A busy switch will be ready to receive dialing at a certain time, regardless of the trunk signaling type. That is, a 'Delay' pulse will be longer than a 'Wink' pulse but both will end at the same time.

A third trunk type doesn't offer acknowledgement to a seizure:

— **'Immediate start'** trunk: the originating end starts dialing a short, fixed time after seizing the trunk. Dial tone from the called switch is optional.

In all cases the called PBX will assert its M lead when the called extension answers. The calling PBX sees this on its E lead.

There is no hard rule against any of these three trunk types having an interface with any of the five types of E&M. However, due to the delay across a frame-based network, it would be difficult, perhaps impossible, for two switches to use delay-dial on any E&M interface. Delay-dial requires a quick response to each seizure, in a time that may be less than the round-trip propagation delay across the frame relay backbone.

Originally, the M lead was toggled between two states (asserted and not asserted) to transmit dial pulses (rotary dialing). By mapping pulse dialing into M lead states, and that into signaling bits, it is possible to transport pulse dialing under the FR Forum Implementation Agreement for VoFR, using the 'dialed digits' subframe payload type.

Signaling between E&M interfaces may involve fairly critical timing of events (see notes in the previous Figure, which is drawn to scale from trunk seizure into dialing). The length of a wink is closely controlled to distinguish it from a disconnect (hang-up) followed immediately by another call request. Replicating dial pulses may have to preserve the ratio of on to off, which varies by country. In the US, the break (open loop circuit, on-hook state) is held close to 61% of the pulse time; in other countries the make:break ratio can be closer to 1:1.

Timing for these older, analog interfaces is the reason the FRF VoFR implementation agreement includes an emulation of ABCD signaling bits. In moving to the future, with ISDN and digital interfaces, the signaling will be based on messages like those in Q.931/Q.933 and Q.SIG from the ITU.

Once the near end has seized a trunk (and been acknowledged by the far end if the trunk type is wink start or delay dial) the originating end will transmit dialing. While it is still possible to toggle the M lead, switches today most often use DTMF (dual tone multi-frequency), the push-button dialing signals. DTMF is much faster than rotary dialing. Between central office switches, there is also a telco system of multi-tone signaling that is not included in the first version of the VoFR IA.

After the called station (telephone) answers (goes 'off hook') the called switch asserts the M signal lead, which appears on the E lead at the calling end. Both E and M remain asserted while the call is in progress. When either switch detects an 'on hook' from its station, it drops its M signal (causing E to drop at the other end). The second switch, if its phone is not hung up, should give it dial tone and behave as if a call were originating.

Note that asserting a signal lead might result in ending a signaling tone on the voice path, as happens with interoffice analog E&M signaling.

Type I: 2 E&M wires, ground

The PBX supplies battery voltage for both E and M in Type I signaling (Fig. 30). Battery on the E lead is presented through a current detector. The positive side of the battery is grounded. The VFRAD asserts the E signal by grounding the E lead at its end, drawing a current that is detected by the PBX.

The transmission equipment (VFRAD) terminates the M lead through a current detector tied to ground. The PBX asserts M by switching its end of the M lead from ground to battery, delivering a current that is detected in the VFRAD.

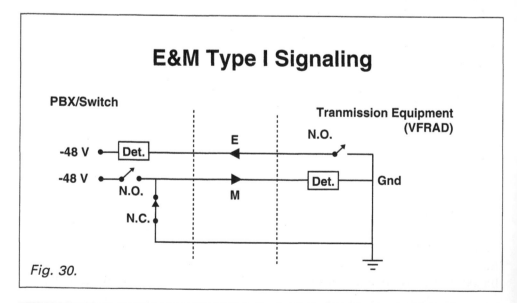

E&M Type I Signaling

Fig. 30.

Type I is common in the US. It is convenient for transmission equipment, like VFRADs, because the PBX supplies battery for both E and M. The circutry in the VFRAD is passive.

Type II: 4 E&M wires

Because the PBX supplies battery on both leads to all trunks in Type I, the unbalanced ground current may become large and cause interference. Type II (Fig. 31) avoids this problem by making both E and M leads balanced. That is:

— The VFRAD does not ground the E lead locally, but at the PBX through a separate lead, sG (signal ground).

— The battery source for the M lead is in the VFRAD, delivered to the PBX on the sB lead (signal battery); the PBX closes the loop to assert M (there is no grounding of M at idle).

Type II is convenient because it is almost symmetrical—except for the location of the detector in the transmission interface being on the grounded leg, where it is on the battery leg in the switch. Still, two Type II devices of the same kind (two switches, or two VFRADs) may be connected together directly (locally) by crossing E to M and M to E (with their signal grounds).

However, Type II also may be inconvenient because it is symmetrical. Now both sides have to supply battery. This is not a problem if there are two real switches connected by 6 or 8 copper wires (2- or 4-wire voice path). However, a VoFR device would need a suitable voltage source (at least -21 V), which is more than the 12 V commonly found in electronics equipment.

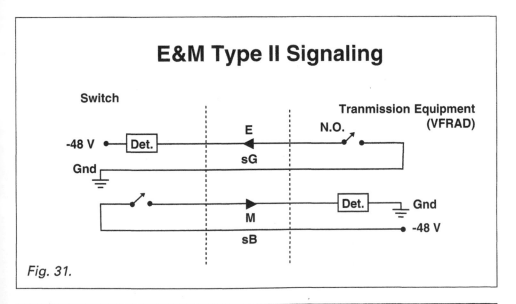

E&M Type II Signaling

Fig. 31.

Type III: 4 E&M wires

The E lead on Type III (Fig. 32) is unbalanced, the same as Type I.

M signaling is the same as Type I (switch from ground to battery to assert) except that both are provided by the VFRAD (via sG and sB leads).

Type IV: 4 E&M wires

The switch side of Type IV (Fig. 33) is identical to the Type II, but the transmission side has the detector in the battery leg of the M circuit. Both sides are identical.

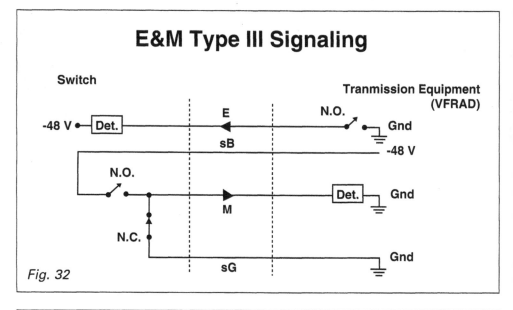

E&M Type III Signaling

Fig. 32

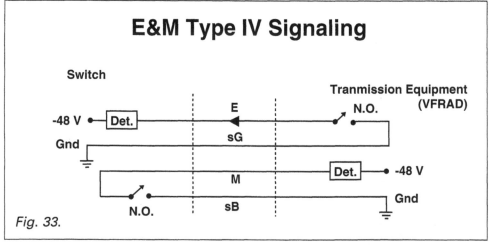

E&M Type IV Signaling

Fig. 33.

Type V: 2 E&M wires, ground

The E lead of Type V (Fig. 34) is the same as Type I.

For the M lead, the VFRAD supplies battery through a current detector. The PBX grounds M to assert a signal, drawing a current that is detected by the VFRAD.

The E lead is symmetrical with the M lead. Thus they may be crossed between similar equipment types to create a back-to-back connection.

Type V is popular in Europe and the UK.

Foreign Exchange (FX) Signaling

Think of FX as an extension cord for the 2-wire analog interface between a switch and a telephone (Fig. 35). When extending a line from a PBX, FX is also called an 'Off-Premise Extension' (OPX).

If this "extension cord" runs through a frame relay network, the voice interface functions must be supplied by VFRADs at each end. The VFRADs mimic a phone at the switch and a switch at the phone. In other words:

1. the 'customer premises equipment interface' (CPE-I) at the customer end of the loop plant must be duplicated in the VFRAD (FXO) on the CO/PBX side of the frame relay link;

2. the switch interface at the CO end of the loop plant must be duplicated in the VFRAD (FXS) on the CPE side of the frame relay link.

At one end of the "cord" (the office end) the switch supplies battery, generates ringing voltage, and accepts dialing. The transmission equipment (the VFRAD) at that end of the FX "extension cord" is 'FXO' for 'foreign exchange, office.'

At the other end of the transmission system, the phone accepts battery current and ringing voltage, and generates dialing. The transmission equipment there (VFRAD) presents an 'FXS' interface ('foreign exchange, subscriber' or 'station') to face the station equipment.

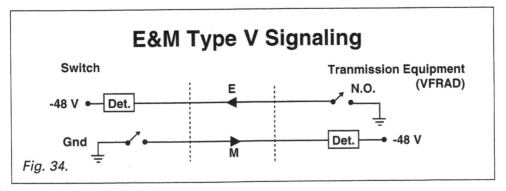

E&M Type V Signaling

Fig. 34.

The name describes what the port faces, not what it is (office or subscriber).

To send a call to a phone over the frame relay network, the PBX rings an extension port as if the phone were attached locally. The local VFRAD, emulating the phone, supports the FXO functions: it absorbs the ringing current, recognizes it as ringing, and sends a call request or "start ringing" message to the remote VFRAD.

When a call request message arrives from the FR network, the remote VFRAD applies ringing voltage to the 2-wire voice interface. When the called phone goes 'off hook' the VFRAD halts ringing and delivers loop current, returning an 'off hook' signal to the calling end. The voice path is established between the VFRADs.

When the far end's 'off hook' message arrives at the originating (calling) VFRAD, it goes 'off hook' itself. This action draws loop current from the PBX and stops ringing. The PBX then opens its talk path to the local VFRAD, which has already set up a talk path to the remote VFRAD. Now, audio energy received by the remote VFRAD on the analog interface is treated as voice: digitized, compressed, packetized, encapsulated in frame relay, and transmitted via the FR network. At the local end the process is reversed to present analog speech to the PBX.

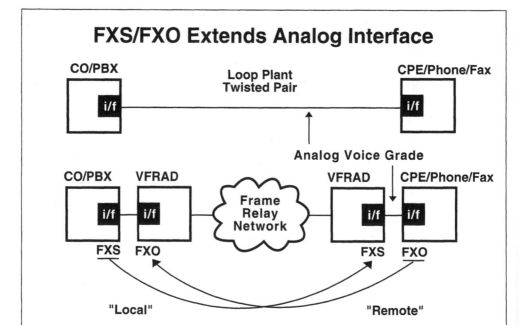

FXS/FXO Extends Analog Interface

Fig. 35. A foreign exchange line simulates the CPE interface (i/f) at the switch or PBX end and simulates the CO interface at the end with the phone. In other words, CO-I = FXS and CPE-I = FXO.

To originate a call at a remote extension, that phone goes off-hook, which draws loop current to tell the remote VFRAD someone wants to place a call. The remote VFRAD signals the local VFRAD (at the switch) via a signaling frame, which uses the same channel ID as the voice path.

The local VFRAD goes 'off hook' to draw loop current from the PBX. This causes the PBX to send dial tone, which is signaled to the remote VFRAD. DTMF or pulse dialing from the remote station arrives as coded messages which are translated by the local VFRAD into re-created DTMF tones or dial pulses. When the remote extension returns to 'on hook' the remote VFRAD signals the local VFRAD to go 'on hook' also, opening the loop to stop drawing current. This indicates to the PBX an 'idle' extension. The call is cleared.

A straight FX function is 2-wire analog at both ends, FXS to FXO. With signaling conversion in the VFRADs, an FX interface (FXS or FXO) may be mixed with an E&M or digital interface at the other end.

On some equipment the voice path is 4-wire (Fig. 36). The two paths are transformer coupled so that each pair may be used as a signaling lead. The

4-Wire Interface for Loop Procedures

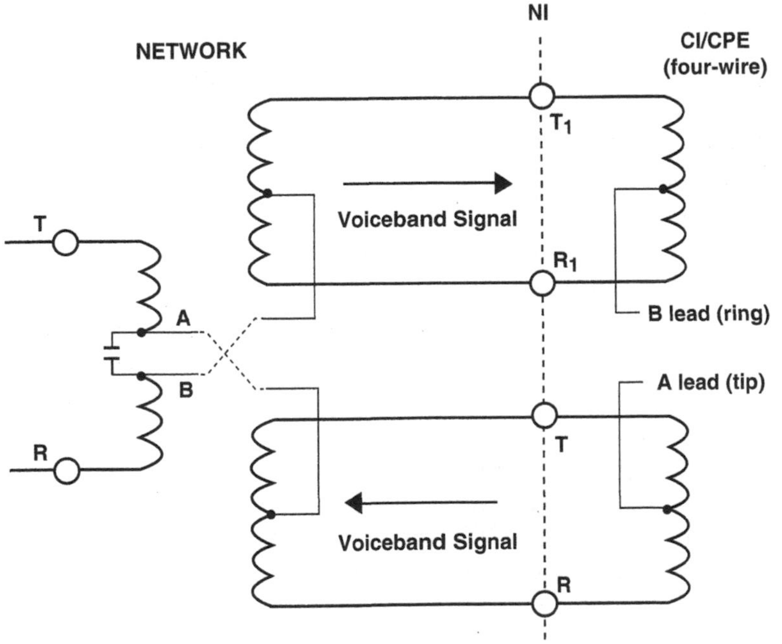

Fig. 36. Loop start procedures are possible on 4-wire interfaces. The two twisted pairs are transformer coupled to allow each pair to act as one signaling lead, taking the place of tip or ring on a 2-wire voice path.

call processing procedures are the same as for a single loop, as described below, substituting the A and B leads for tip and ring.

With a 4-wire interface, a signal in each direction feeds directly into an amplifier where necessary. A hybrid is not needed to split the signal—eliminating the prime source of echo. Lack of echo is one of the advantages of this interface type, particularly when the network is 4-wire also.

With all that said, there are two ways to use the analog loop for supervision of on- or off-hook status, 'Loop Start' and 'Ground Start.'

Loop Start

The standard analog telephone line in a residence or business is 'loop start.' A PBX supplies the same interface to analog extensions.

A simplified loop start circuit (Fig. 37) shows the essential elements. A battery in the CO operates the switch and provides power to the phone.

The loop start line may be used to send or receive calls at a telephone or a manual PBX. If terminated by an automatic PBX or another switch, a loop start line must be restricted to either incoming or outgoing calls. Trying to do both automatically leads to a conflict called 'glare' (see Index).

Calling Procedures

When the line is idle, the phone is on-hook and the hook switch is open. The capacitor (C) in series with the ringing detector does not pass d.c. current (from the battery). With no d.c. path through the phone, the loop current through the loop plant is near zero.

To initiate a call, the caller picks up the phone (goes off-hook), closing the hook switch and letting the speech network (hybrid) draw current from the battery. This current powers the phone.

The CO switch sees this loop current as a request for service and a notice to prepare to receive a dialed phone number. When ready, usually within 3 seconds, the switch gives dial tone and no more than 70 ms later is ready for the caller to dial.

During dialing and call processing, depending on the make of the switch, the voltage from the network may reverse polarity, change up or down, or disappear briefly. If the CPE opens the loop during dialing for more than 1 ms, that may be counted as a pulse; an open circuit of more than 100 ms may disconnect the call.

Audible call progress signals are usually given to the caller: dial tone first; then ringing tone to indicate the far end has not yet answered, busy signal, or re-order (fast busy). At the end of the call, the phone is hung up (goes

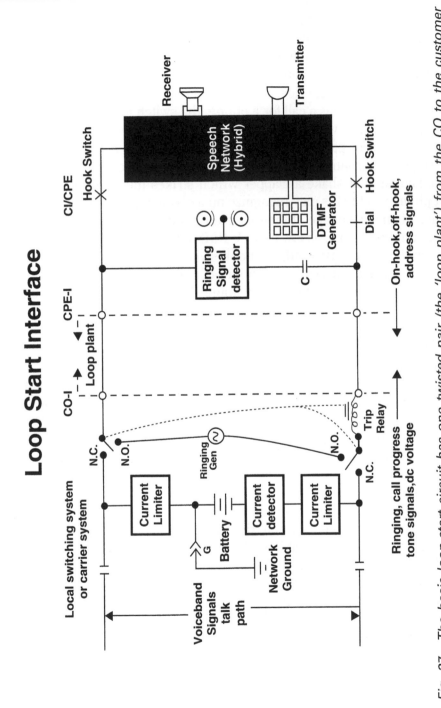

Fig. 37. The basic loop-start circuit has one twisted pair (the 'loop plant') from the CO to the customer premises equipment. The CO and the CPE look at the loop with interfaces labeled CO-I and CPE-I.

on-hook) which stops the loop current. An open loop tells the switch the line is idle again and to clear the call.

On lines with extra features that are activated by a 'hook flash' like call transfer, call waiting, or 3-way calling, any on-hook period over 1.5 s is considered a disconnect. An open interval of 300 to 1000 ms is taken as a hook flash signal to invoke a service.

When the switch has a call for a phone, the switch alerts the called party by applying an a.c. voltage (up to 170 V a.c. RMS) to the local loop. Alternating current passes through the capacitor C to activate the bell or other ringer. Alternating current into a standard analog phone energizes electromagnets that shake a clapper which strikes mechanical bells. More recent phones and PBXs may recognize an a.c. ringing current or merely the presence of an a.c. voltage of the expected frequency and amplitude (above some minimum voltage, as low as 40 V RMS for a 20 Hz ringer).

When the call is answered, the hook switch closes the d.c. loop. Battery current trips a relay or electronic monitor in the switch, which disconnects the line from the source of ringing voltage (in under 200 ms) and connects it to the talk path.

VFRAD Requirements

The loop start FXS interface provides an analog voice grade connection suitable for a standard '2500 telephone' set, key system, or Fax machine. The VFRAD should provide "standard" battery and ringing voltage. The choice of the standard is left to the vendor. Some choose Pub. 43801, the original description for the CO channel bank. Others choose Bellcore's TR-303 (or the slightly older TR-57) for shorter loops from remote digital terminals designed to be placed in cabinets at the end of the street. The ANSI specification for PBX interfaces is another option.

The VFRAD that offers loop start FXS support must deliver ringing voltage to the called equipment when 'ringing' is signaled from the frame relay network. Ringing should be at least 40 V rms for a load of five 'ringer equivalents' (REN = 5) REN = 1 is 8000 ohms, so five in parallel is 1600 ohms.

Careful choice of applications could allow lower voltage or lower power if the phone responds to ringing current. However, some PBXs draw practically no current: they respond only to ringing voltage near the CO standard source voltage of about 100 V.

Standard ringing cadence is 2 s on, 4 s off. Other "on" and "off" periods should be programmable, particularly when there are more than three analog interfaces on a VFRAD. To save on the cost of the ringing generator, only one phone should be rung at a time. This could mean 1.5 seconds in 6, for example, when there are four voice ports.

A signal frame or other indication should be returned to the calling end so the VFRAD there can generate audible ringing tone for the caller.

The battery voltage could be 24 V, or even less, if the phone equipment is close to the VFRAD. Again, some PBXs may be sensitive to this voltage.

The FXO requirements are nowhere as critical, because this is the passive side of the interface. The key requirement is the ability to withstand full ringing voltage from a PBX on top of the full battery voltage.

Ground Start

An examination of the loop-start process reveals a problem if it is used on a two-way interface between two switches, as when a PBX or VFRAD is connected to central office lines. After a loop-start line has been seized by the calling switch, there is a period of up to 4 seconds before the calling switch applies ringing voltage for the first time. During those 4 seconds the called switch doesn't know that the line is in use and could attempt to make a call on the same trunk. This condition, called 'glare,' prevents the line from being used for either call attempt. Glare is important when both sides are automatic; loop start can be used for 2-way calling if one side is operated by an attendant who can deal with the condition.

To avoid glare, the 'ground start' (GS) interface is used between switches on two-way trunks. It has a more positive process to seize the line: one of the leads is grounded; ringing may be optional. The CPE side grounds the Ring lead, the network side grounds the Tip lead. In this context, the CPE is the PBX facing the VFRAD, which acts like network.

If you look back at E&M-IV signaling, you'll see a strong resemblance to the basic ground start circuit (Fig. 38). Each side applies battery, through a current detector, to a lead that the other side may ground. Unlike the E&M interface, GS is not fully symmetrical. Each side of the interface (FXS or FXO) is most often supplied in a different module, especially when ringing voltage is to be supplied by the FXS.

In the ground-start idle state the VFRAD's battery (+ side) is grounded by contact G. Contacts B and S, the ones used to ground the tip and ring leads, are open.

The current detector on each side of the interface registers when the other side grounds the lead and draws current from the battery. This ground signal is applied at seizure, so the called switch knows immediately that this line is not available. Grounding a lead is a positive indication of a new call; ringing voltage is not absolutely essential.

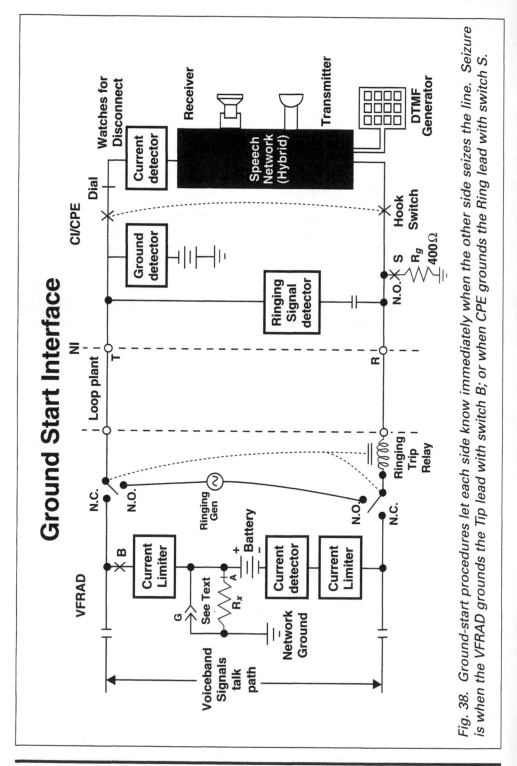

Fig. 38. Ground-start procedures let each side know immediately when the other side seizes the line. Seizure is when the VFRAD grounds the Tip lead with switch B; or when CPE grounds the Ring lead with switch S.

Calling Procedure

The procedure for the customer to originate a call is to close contact S, grounding the ring lead. At the same time the hookswitch places the phone circuit on the loop (tip and ring leads). Contact S draws current through the ring lead only, from the battery in the network side (in the VFRAD).

When the FXS VFRAD port (like the CO switch) recognizes the request for service, and is ready to receive dialed digits, it closes contact B to supply loop current and may open contact A. It may supply dial tone. The caller then dials the called number.

If the circuit has what's called 'floating battery,' the more modern type, contact A is closed only during idle. This ground completes the circuit with the ring lead so contact S can draw current. 'Conventional battery' in the VFRAD or switch is grounded at all times: contact G is always closed. Only one switch will exist in any circuit Interface.

When the FXS VFRAD port or switch originates the connection on the interface it is 'terminating a call' (because the call originated somewhere else). The alerting action is to close contact B, grounding the tip lead, and apply ringing voltage to the ring lead (on top of the battery voltage). Note that the "2 s on, 4 s off" cadence of ringing may mean that the tip line is grounded for some seconds before a.c. ringing voltage appears.

The PBX may answer immediately, on sensing the grounded tip lead, or it may react only to ringing voltage. In either case, the PBX (FXO port) answers by going on-hook. The loop current trips the ringing generator off, cuts through the talk path, and opens the battery grounding contact A (if using a floating battery).

After recognizing a new call, the GS interface operates like loop-start until both sides go back on-hook. The disconnect process depends on which side goes first and whether the battery is floating or conventional.

If the terminal (PBX) disconnects, it opens the loop by going on-hook (breaking the contacts of the hookswitch). If it is a conventional battery on the other side of the interface (in the FXS VFRAD or CO switch), the PBX waits until ground is removed from the tip lead, then a short guard time, before returning the line to idle.

If the switch or FXS VFRAD disconnects first, it opens contact B, stopping the loop current. A floating battery circuit will also disconnect the negative battery side from the ring lead for about half a second. The current detector in the PBX waits 350 ms to confirm that the open loop is not just a transient, then may wait a short guard time or immediately return the port to idle.

Ground start has other useful features that weren't imagined when it was introduced in the 1920s:

— The ground is removed by each side when it returns the line to idle. Positive indication of a disconnection before each new call helps automatic call distributors deal with high volumes of calls into hunt groups.

— Disconnect supervision prevents certain types of call fraud.

— Each end can tell when the other end has hung up.

In England "ground" is called "earth" which leads to "earth calling" as the name for this interface type. But they've already heard the joke about "Earth calling Flash Gordon."

Potential Problems

Some central office switches test each loop before or after making a connection to it. The tests look for a power cross (high voltage on the loop that could damage the switch), shorts that would immediately trip the ringing circuit, and other conditions. Test procedures usually involve applying battery and ground to tip and ring, usually through resistors, or attaching resistors alone to the leads and looking for current flow.

These tests are very brief, as short a 10 ms and seldom longer than 50-100 ms. Yet problems arise with some ground start terminal equipment when it originates a call. An attached resistance applied between the line seizure and the full response can be taken as a proper ground. The terminal then starts dialing before the switch is ready to receive the digits. Brief open-loop conditions, as the switch moves the leads from idle condition to test equipment and back, may be seen as a disconnect.

Not all switches impose these transients, but most of the equipment installed in North America does.

Another class of problem is seen if the VFRAD's FXO port is facing an older switch, for example a crossbar or step-by-step (now almost gone). These mechanical switches may disconnect a call, go the idle state, and seize the line again in less than 50 ms. To most terminal equipment (and probably the VFRAD) this looks like a blip during a call—not a new call. The result is 'ring no answer' where the caller hears a ringback tone but nobody ever answers because the called equipment doesn't recognize a new call.

Newer electronic switching systems wait for at least 600 ms, a guard time that ensures the FXO VFRAD port will see the disconnect and a new call. It helps if the terminal device being called looks for ringing voltage as well as the grounded lead.

Most CO switches test for low-resistance loops before completing a call. If the terminal (PBX or FXO port on a VFRAD) has not yet gone on-hook, the loop will have low resistance. Connecting to the terminal port would immediately trip ringing off, so if the terminal needed ringing to detect a new call, that would not happen. The result is 'no ring no answer,' which the test tries to prevent. A danger lies in putting a low resistance across a line in a hunt group to busy it out. If that line is at the preferred end of the hunt pattern, it will fail the test and the switch will try the next line in the hunt group. If that one passes and the call completes, OK. If the second line is busied out also, the CO switch will give the caller reorder tone and clear the call. As few as three or four ground start lines in a hunt group busied with resistors will make almost all incoming calls attempts fail.

Automatic Ringdown (ARD)

The foreign exchange line, described just above, incorporates the essential feature of automatic ring down: going off-hook at one end produces an immediate and pre-programmed action at the far end of the transmission line. In the case of FX, the action is to mimic a copper cable.

ARD, which also goes by the name 'private line ARD' (PLARD), has a different action. It links two phones so that when one goes 'off hook' the other rings immediately. Neither end needs to dial anything to cause ringing at the far end. The phones may have no dials.

Airports and other travel-related sites offer instant access to reservations, taxis, restaurants, and many other services simply by picking up the designated phone. That's and ARD line.

The detailed function is essentially that of the foreign exchange line described in the previous section, but with FXS interfaces at both ends. The calling end still sends an "off-hook" message when the phone is picked up. But the called FXS reacts with ringing where the FXO would have responded with off-hook (drawing loop current). Most ARD implementations require both phones be returned to 'on hook' before ringing can occur again.

FXS interfaces are needed at both ends of the transmission equipment because they supply ringing. FXS is available in almost all VFRADs with analog voice ports. Not all VFRADs, however, support the signaling configuration which produces the ARD function.

A variant is called 'manual ringdown.' The MRD voice path is fixed between two FXS interfaces, the same as ARD, but ringing occurs only when the "caller" presses a button. The button applies ringing voltage to the far phone, regardless of whether the caller is on or off hook.

Traditionally the connection is for short distances only, where a pure copper connection is available between the two phones. All known users are securities traders in New York City. They like the ability to use ringing patterns to get special attention—three short rings means "hot deal."

VoFR doesn't define MRD explicitly, but the feature should be programmable (by the vendor) using the DTMF signaling syntax in a special way on a specific subframe channel. The "magic button" could be defined as the * or # key, or the famous "Any" key (as in "press any key").

Digital Voice Interfaces

Digital voice circuits with Channel Associated Signaling (CAS) already have signaling information in the form of ABCD bits. These are bits in a DS-1 frame that are reserved for signaling status of a particular channel. These bits are positioned in a T-1 or E-1 channel in a way to distinguish them from encoded voice. The channel bank defined the format in the domain of time division multiplexing:

— the "D4" format for T-1 lines carries ABCD as "robbed bits" in each time slot (lsb's) every sixth frame of the superframe,

— "CAS signaling" in time slot 16 (which is the 17th in the frame because the first is "number 0"). TS16, though only 8 bits per frame, repeats 16 times within a superframe to provide a unique space for the four ABCD signaling bits associated with each voice channel.

Both T-1 and E-1 are said to have 'channel associated signaling' because certain bits embedded in the overall voice signal are devoted to each voice channel. It is also called 'robbed bit signaling' on a T-1 line because one bit in 48 is "robbed" from the voice channel to convey signaling.

By contrast, the ISDN Primary Rate Interface (PRI) reserves one time slot (#24 in a T-1 or "16" in an E-1) to carry messages for all channels. There are no reserved bit positions. Since the time slot is shared, this form is called Common Channel Signaling (CCS). (See the author's *Guide to T1 Networking* and *ISDN User's Guide* for details.)

The D4 format applies to other equipment with T-1 interfaces as well as channel banks because the standard evolved for the long-haul portion of the circuit. Even small digital PBXs offer T-1 or E-1 interfaces.

On frame relay networks, various vendors have solved these signaling problems, but each in its own way. At this writing, there is no interoperability between vendors for voice carried over frame relay networks (VoFR). Publication of the final VoFR IA is expected to change that.

Ironically, the transport method proposed for CAS signaling bits (in Annex B of the IA) again mimics the local loop. A VFRAD captures what happens on an analog voice interface by converting loop activity to four signaling bits, A, B, C, and D. The result should be the same ABCD bit activity a channel bank would create if the analog interface were being converted to digital form on a T-1 line, though the meaning of the bits is not part of the IA.

Fortunately, a digital interface with CAS has these bits present. A VFRAD simply has to collect and transmit them, as described earlier. Since this is relatively straightforward, digital interfaces may be the first to achieve interoperability over frame relay.

For digital voice interfaces with common channel signaling, initial releases of VoFR equipment will simply carry the signaling messages across the FR network. Signaling frames are in an HDLC format that is easily encapsulated in VoFR subframes for transport.

ISDN interfaces on digital PBXs use various forms of messages defined by ANSI in North America as Digital Subscriber Signaling System 1, DSS-1 (wonder what happened to the other S), and in Europe by ITU with a standard under development known as Q.sig.

It will require considerably more work to define a standard way to terminate these signaling links on a VFRAD which interprets and acts on the messages. In this capacity the VFRAD will have to perform many of the functions of a signaling transfer point. The STP is a specialized packet switch used in the public networks to handle common channel signaling messages. Some functionality like an STP will be necessary to fulfill the expectations of major corporations who want to implement dialing plans over the VoFR network, using SVC service on the backbone.

Dialing via VFRAD Over PVCs

This sounds contradictory. It's not. The PVCs link the VFRADs, while the dialed connections are set up among the VFRADs, between individual voice ports in a network. The VFRADs use the frame relay PVCs as trunks. Individual calls are multiplexed on one DLCI with additional addresses in the subframe header.

The main question on dialing is whether the VFRAD simply passes dialing or makes decisions based on the dialed digits. Transparent transmission is easy to configure and generic, even standardized as the FX feature. Interpretation, so far, is proprietary.

Transmission Only

When mentioned with the FX description earlier, dialing came after the voice path was established between the calling phone at a remote site and the central PBX. DTMF dialing was carried transparently from the phone to the switch (PBX). The VFRAD understood the DTMF tones only to the extent needed to convert them to special codes in the calling VFRAD (for transmission in signaling subframes) and reproduce those tones in the central VFRAD for the PBX. The PBX did all call processing based on dialed numbers.

Rotary or pulse dialing can be handled within the CAS signaling transfer syntax. The opened and closed status of the interrupter switch in the remote phone is transported and reproduced by the central VFRAD the same as the hook switch position—just faster. Every break or pulse sent by the rotary dial phone is reflected as a brief opening of the hook switch in the FXO interface.

Interpretation/Routing

A VFRAD that can compress voice contains a powerful DSP that monitors the voice interface. All VFRADs also have a microprocessor to handle protocols and other functions. This amount of processing power is capable of interpreting dialed digits and performing a wide variety of functions based on the called number. "All it takes is software."

The VFRAD's position is quite different when configured to interpret dialing. In the FX case the PBX provides dial tone over the frame relay network. But when the VFRAD interprets dialed digits before setting up the connection, dial tone must come from the VFRAD when an attached phone goes off-hook.

Only a handful of VFRAD vendors have gotten very far with dialing. The best versions at this writing offer some very flexible features for constructing private voice networks on frame relay backbones. Depending on the network design, a PBX or voice switch may not be needed.

Call Routing

The simplest feature, but very useful, is to map a call request to a specific frame relay connection (PVC). The VFRAD maintains a routing table which matches called numbers to DLCIs. When the VFRAD interprets a dial sequence as a called phone number, it looks in the routing table to see which frame relay connection to use.

This feature allows a caller to select the FX link used for each call. This is a way to access multiple PBXs in different locations.

Vendors who get this far usually provide the ability to map a call down to a specific port at the called site, using subframe addresses.

Hunt Groups

If the specific voice port at a called site is not critical, the VFRAD may map one called number to any available voice port in a hunt group at that site. That is, a caller who wants tech support dials one number to reach any of the six people at the help desk.

A hunt group is also called a rotary. If any port in the hunt group is free, a caller gets connected. If all of the ports in the hunt group are busy, the caller gets a busy signal.

Outward Dialing

The calling VFRAD maps call requests to different DLCIs which may terminate at phones or at PBXs. To call a phone is simple: the phone rings. To call a PBX may be as simple, if the PBX is programmed for ARD on that port, for example, or the PBX returns a second dial tone to the caller.

There is another choice that offers more flexibility: have the called VFRAD 'place a call' on the PBX. That is, the called VFRAD (connected to the PBX) can be configured to dial another phone number into the PBX to complete the call. The number might be associated with the port on the called VFRAD, or it might come from the calling VFRAD as part of the call request.

Vendors with voice interfaces that can be certified for connection to the PSTN offer an interesting option to long distance. The outward dialing can be over the PSTN, to any destination. Usually it is corporations with large frame relay networks, private or based on public carriers, who use that network for the long distance portion of calls. Going out to the PSTN for call completion is often just a local call. This form of private/public interconnection is prohibited by some PTTs, usually those that charge exorbitant rates for international calls. However, now that international call-back is legal in many places, the pressure from frame relay (and some day the Internet) will eventually make VoFR a common option. Frame relay SVC service will push harder in this direction.

Call Switching

The calling VFRAD may not have a pre-configured PVC to the VFRAD at the called phone's site. This happens in the most common network topology, a star, where each remote site has a PVC to the central site only. In fact, a full mesh of PVCs is most often prohibitively expensive on public frame relay networks. For one remote site to reach another remote site, the connection must pass through the central site.

With an FX connection between the central VFRAD and each remote VFRAD, the call switching is done by the PBX. The drawback is that one call occupies two voice ports on the PBX and two on the central VFRAD. Switching software in the central VFRAD can route calls between remote sites without involving the PBX.

Switching works in conjunction with the dialing interpretation feature. The calling VFRAD, having understood the called number, looks in the routing table and finds the site where that number is assigned. The calling

VFRAD tells the central VFRAD, in the call request, to switch the call to the other remote site in frame relay format. The central VFRAD then acts as a frame relay switch for this call connection. That is, voice frames arriving at the central VFRAD are returned to the FR network with a different DLCI (the one for the PVC to the called site).

Switching of Frames & Subframes

Packet switching allows consecutive voice frames to be separated. The frame relay network deals with each frame individually. That is, consecutive voice frames usually aren't sent contiguously on the transmission line. Between voice frames, data frames could be interspersed, or an idle condition. In fact, as seen when we considered compressed voice in frames, some time will elapse between the transmission of consecutive voice frames that constitute one connection. This time interval lets data or other voice channels use the frame relay access line.

The private network or public frame relay service routes a frame to the proper destination indicated by the address in the frame relay header. Dealing with frames this way inherently allows for statistical multiplexing over lines and physical interfaces. That is, the network can mix voice and data and let many people share each access link if each user has a different DLCI.

In today's networks, a frame relay permanent virtual circuit (PVC) has just two ends. While the FR Forum defined multicast or broadcast services in October 1994, there is no such service available now from a public carrier. To date, when many conversations share one frame relay DLCI (using subframes with channel identifications, CIDs), they all originate and terminate on the same two devices (pieces of customer premises equipment). One device is at each end of the virtual circuit (VC). These devices might be PBXs, VFRADs, or other network access equipment.

This 2-point requirement does not prevent an access device from terminating more than one frame relay VC, nor more than one conversation per VC (using the CID). But each *frame relay* address has to go to just one place.

That place could be a gateway device. Some day this might be carrier's equipment inside the network. Today it is CPE. Hardware that is both gateway and switch can route submultiplexed connections from one VC (DLCI) to another VC by changing the DLCI and CID. Several models of this gateway exist, using proprietary formats. They work with an address in a subframe header, using its value to help make routing decisions and possibly to change that subaddress as the frame passes through the gateway (Fig. 39).

This channel-switching feature can route a call through an intermediate VFRAD node (phone B connected to phone D) in compressed digital form.

When a frame arrives, the VFRAD knows its identity from the port it arrived on, the DLCI, and the CID. Those numbers point to an entry in a routing table. A look-up in the table produces the handling instructions: whether to terminate the frame locally (and on which port), or forward it to a port ID, DLCI, and CID. The VFRAD does not have to convert the voice information to analog or even decompress it.

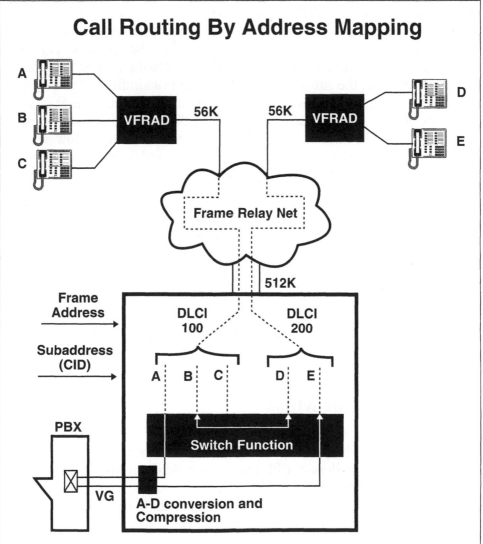

Call Routing By Address Mapping

Fig. 39. Existing CPE manipulates the frame relay address and the sub-header address to route voice frames between VFRADs. Path may be determined by manual configuration or on-demand by interpreting dialed digits from the calling VFRAD.

Note that this drawing shows the two PVCs on the same access link—they could be on different links directed at different networks (one public and one private, for example). The connection from phone A to phone E is routed through the PBX, as would be needed if the FRAD did not also switch frames.

The FR address (DLCI) is said to have only 'local significance.' This means that the address applies to a logical connection only on a physical link between devices. The address on a voice frame may be changed as the frame passes through any FR switch, including a gateway. The DLCI of a logical connection can take on many values at different points in the backbone network. In this regard voice frames again are identical to data frames. The DLCI assigned to a connection directs individual frames to the proper VoFR access device or terminal; the sub-address or CID targets an individual voice port.

If the gateway were to implement the multicast function, that is, make duplicate frames and forward the copies to different addresses, then the gateway would also emulate a multidrop line or voice bridge. However, this multicast function is not a full telephone conference bridge. Those are much more sophisticated and may need a DSP for echo cancellation and switching among speakers.

Looking Toward SVCs

No public frame relay service offers switched virtual circuits (SVCs), so far. There seems to be some reluctance among carriers to cannibalize their existing voice revenues, which is a justifiable expectation. Announced dates to introduce SVCs have slipped repeatedly, for years.

Meanwhile, frame relay switch makers are pushing to develop the new software that will allow installed switches to support SVC services. Most of the major vendors have demonstrated or at least announced SVC features.

Big Benefits for Most

Benefits of SVCs to the carries can be huge. Their biggest problem today with frame relay is administering the DLCI assignments, which is a manual process. Installers of CPE complain most often that they don't have the right DLCI information.

Users of the service have similar trouble tracking DLCIs. Most frame relay equipment installed today is configured manually for the frame relay parameters.

Dial backup protection against the loss of a frame relay access line is complicated by the fact that frame relay networks support only PVCs. If a remote site calls into the host site, how is that equipment to know which

PVC to swap over from the backbone to the new caller? There are ways, but most are complex.

All of these DLCI problems disappear with SVCs. At present the gain is still perceived to be less than the loss of revenue expected when voice calls migrate from the PSTN to the frame relay backbone.

When one carrier has installed the new software, and finished testing. SVCs will follow in a flood.

Big Impact on VoFR

The goal of the VoFR Implementation Agreement is to allow every VoFR unit to communicate with any other. That is, large and small devices from one vendor should network with themselves or with other brands over permanent virtual circuits (PVCs). This is not likely soon, even if the IA is adopted in 1997, but it will happen.

Changes will be needed to update the VoFR implementation agreement to include SVCs. SVC service means that a user no longer can rely on having the same type or brand of VoFR equipment at both ends of a connection. In fact, there is no assurance that both ends will prefer the same compression algorithm.

In an SVC environment, VoFR FRADs will have to negotiate the connection parameters for each call. Which algorithm? What speed? What kind of signaling? Support Fax? At what speed?

When switched virtual circuits (SVCs) become available, the pressure to interoperate between enterprises will lead vendors to adopt a second phase IA that defines the additional functions for voice needed by end users of SVCs.

Much of the negotiation procedure needed for SVCs was discussed in the FRF technical committee around 1996. These procedures were dropped from the first edition of the IA not only because SVCs weren't available, but also to speed publication. Putting that material back in should take less work (and less time) than is typical for a major revision of an Implementation Agreement.

Eventually, all will be worked out. The result should be vastly increased popularity of VoFR, frame relay services, and digital communications in general. Growth in VoFR (most optimistic case) could rival the growth in facsimile machines after the Group III standard was adopted.

Chapter 5

Applications of Voice & Data

A Voice over Frame Relay (VoFR) system provides voice channels on a frame relay data network. The voice quality is more than acceptable for connections within an enterprise, earning Mean Opinion Scores comparable to ADPCM (MOS of 3.5 to 4).

Technically, frame relay excels in having relatively low transmission delay, which is nearly constant compared to other transmission types, notably X.25 and TCP/IP. The error rate is so low that frames are lost only rarely. The high quality is due in large part to the optical fiber cables in the US public FR networks. The same cables carry the leased circuits that private networks are based on.

With frame relay services spreading worldwide, more networks have a global reach. Adding voice to a global data network can show a payback measured in weeks, compared to separate leased voice lines or international dialing.

Up to a dozen sources of VoFR equipment offer a wide range of equipment to fit many network applications. Many more vendors supply frame relay equipment for data-only applications.

In the short term, don't ignore the example of early packet voice servers. A single-vendor solution that meets your needs will save serious money in a large network. Just two VoFR devices linked by a leased line or frame relay service, which ever is less expensive, can save on phone bills between two points. You don't need a huge network to justify VoFR.

Key Application Examples

The focus here is on the voice component. Data connections will be shown but not discussed in detail. For clarity, voice concepts are described in isolation. A practical network will combine many of these features.

Each network must be configured to meet unique needs, which are certain to change over time. The specific parameters given in the examples here are just that, examples. None of the details should be taken as an absolute requirement. Where economics or convenience dictate, voice interfaces may be digital as easily as analog without affecting the principles involved in the examples.

PBX Tie Trunks

Customers often prefer E&M connections as a PBX interface. E&M is standard on some PBXs for trunks to the CO as well as tie trunks to other PBXs.

Pulse dialing is still used. Once the only dialing type, for many years pulse dialing was less expensive than DTMF service from the telcos. PBXs saved money by converting DTMF from the user to pulse on the CO trunk. Now some tariffs charge the same for either pulse or the much faster DTMF.

If moving some trunk PBX ports from switched telco service to VoFR equipment (Fig. 40), you will have to gather information on the configuration. Note if the PBX dialing is pulse. If so, consider changing it to DTMF while installing the VFRAD. Calls will connect noticeably faster.

Automatic Ringdown

A cash dispenser (automated teller machine, the "other" ATM) in a grocery store is placed and maintained by a bank. The store merely provides

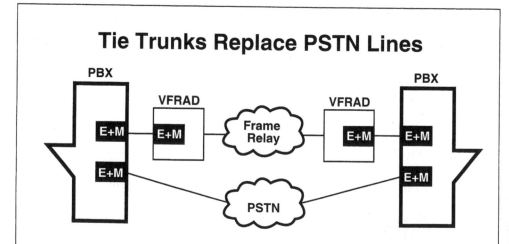

Tie Trunks Replace PSTN Lines

Fig. 40. When call volume is large enough, tie trunks based on VoFR will save money compared to dialing on the Public Switched Telephone Network.

the space. If the machine eats your card, to whom do you take the problem? Not the store manager, he isn't involved.

That's why some banks provide a phone on the cash machine: "To speak with a customer service representative, pick up this phone." There is no dial needed—all calls go to only one destination. Lifting the handset automatically rings a phone in the customer service office using the VoFR network to implement automatic ringdown (ARD) (Fig. 41).

The ATM and the phone share a VFRAD that has only one link to the frame relay network, reducing costs to the bank and minimizing the amount of wiring needed in the store.

Both phones may connect physically to FXS ports on VFRADs, though the service center is likely to have a PBX for call distribution. If so, the ARD service will involve other features, described below.

While possible, it is unlikely that a call will be placed to the phone at the ATM site.

Foreign Exchange

A large organization with many branch offices might need voice and fax connections to each remote location. To let a branch office telephone appear to be an extension off the PBX at headquarters, the foreign exchange con-

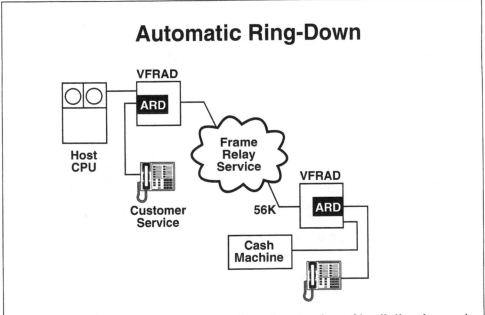

Fig. 41. Lifting one phone causes the other to ring. No dialing is needed or possible. The connection is predefined between two points only.

figuration on the VFRADs (Fig. 42) acts as an "extension cord" for the phone to reach the PBX.

The remote VFRAD has an FXS interface to work into the phone. The central VFRAD has an FXO interface to work with the switch. Details of their functions are given in Chapter 4.

When the phone is lifted, the frame relay connection is set up automatically between the phone and the switch. The process is similar to ARD, but the FXO draws loop current where the ARD applies ringing.

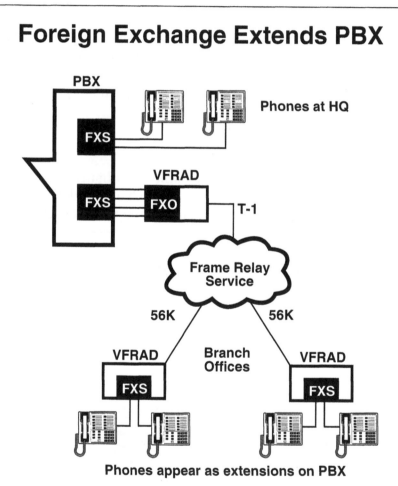

Foreign Exchange Extends PBX

Phones appear as extensions on PBX

Fig. 42. When branches and headquarters call each other often, it's convenient to have all the phones work like extensions on the same PBX. The FX feature allows a port on the PBX to reach a phone anywhere on the FR network. Other features—hunt group and call routing—let many remote phones operate through far fewer PBX ports, reducing costs of equipment at headquarters.

Depending on what is attached to the far end of an FX, the caller may receive any of several possible progress signals:

— Dial tone, indicating connection to a switch that will accept pulse or DTMF dialing instructions;

— Ring-back tone, indicating a call attempt to a pre-programmed destination or another phone (ARD);

— Busy, if the far end of the PVC is unavailable,

— Fast busy, indicating the PVC is "inactive" or "not present" as indicated by the LMI or Annex D status report.

The dial tone heard by the caller comes from the PBX. Dialing is carried transparently from phone to PBX. It is the PBX that routes the call, generates call progress tones (like audible ringing for the caller), and may provide other PBX features, including PSTN access.

FX applications may use a VFRAD feature like line contention or route selection, described in sections below.

Line Contention/Hunt Group

For large banks with many ATMs, or insurance companies with many branch offices, it is not practical to have a separate phone instrument or PBX port at headquarters for every phone in a remote site. In fact, the number of connections should be limited to the number of people available to answer calls. For some troubleshooting services, the calls from any remote site are infrequent and of short duration. In practice, the number of calls that have to be supported at any given time may be much smaller than the number of phones installed.

For many reasons there may not be a need for as many phones at HQ as there are remote phones. But then there is a need for all remote phones to share the HQ agents. That is, when a branch wants to talk, that branch must be able to use any open phone at headquarters. Somehow, the network must be able to distribute calls from many remote sites over a smaller number of voice ports at HQ.

Port contention is the solution for a frame relay network, implemented in VFRADs. The network diagram is the same as the previous figure. The key function is in the central site VFRAD, typically a large unit with as many as 60 voice ports. This central VFRAD will have a large bandwidth link (or links) to the backbone, at least one T-1 or E-1, two links for improved reliability.

Hundreds of virtual circuits from branch offices can be provisioned on each frame relay link into headquarters. The large VFRAD there directs each connection request to the next available voice port. As long as any voice port is idle, incoming connection requests are satisfied. If all the ports are busy, the caller hears a busy signal generated by the remote VFRAD when the central VFRAD refuses the connection request.

If the number of agents ready to answer calls drops below the number of ports on the central VFRAD, the network operator can busy out the extra ports. Callers then get an answer or a busy, not a ring/no-answer.

Calls made from the HQ site to the branches use the call routing feature, described next.

Call Routing at Origination

If there are multiple possible destinations for the call (not ARD), some device will have to switch it. Most often in a private network this is a company-owned PBX or other voice switch. In a VoFR application, the PBX could do the outbound routing from headquarters (see previous example) if there were a separate PBX port for each remote site. Lately, VFRADs have displayed the ability to interpret dialed digits and route a call (Fig. 43).

To act on the dialed number, the VFRAD must process the information, not just pass it transparently to another device (like a PBX). In other words, the VFRAD deals with a phone during the dialing interval. When the phone goes off-hook, it is the VFRAD that gives dial tone, not a remote PBX.

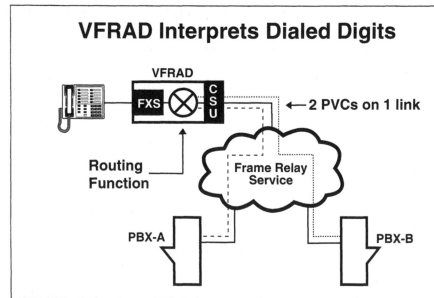

VFRAD Interprets Dialed Digits

← 2 PVCs on 1 link

Fig. 43. To capture dialed digits, the VFRAD must be the source of dial tone. After dialing, the number is looked up in the routing table to determine the destination PVC number.

The caller dials as usual—there is no way to tell the source of the dial tone. The dialed digits ('rotary' pulse trains or DTMF tones) are held by the VFRAD until dialing is completed. The VFRAD then analyzes the number to see if it can find the destination. One vendor stores a complete map of the voice network as a routing table in every VFRAD. If the number is in the table, it means the called phone is reachable from this VFRAD.

The routing table in the originating VFRAD identifies which frame relay port and which DLCI to use in the next step: to send a connection request message to another VFRAD, which accepts the connection and completes the call on a voice port. The final action, alerting the called party, depends on the type of port as described in Chapter 4.

If the called phone is not attached locally, the VFRAD will have to pass on the call request, which it can do in at least two ways:

1. Outward dialing; the VFRAD puts the call request on a local voice port attached to a PBX or CO trunk. When that switch gives dial tone, the VFRAD dials another number (from the routing table or in the call request). That number may be an internal extension or a PSTN call anywhere in the world.

2. The VFRAD performs digital switching, as described in the next example.

Voice Hub with Switching

Almost all of the early VoFR applications were on networks with a star topology, partly of necessity. The assumption was that the branches spoke only with headquarters, not with each other.

It turned out that there are frequent occasions when one branch needs to talk with another branch. Bank customers may go to branches that do not have their signature cards on file, for example.

If the network is based on foreign exchange service, as described above, then branch-to-branch connections pass through the PBX. Consider that one such call ties up two ports on the VFRAD as well as two ports on the PBX. If the volume of branch-to-branch calls is significant, the burden on the PBX may be unacceptable. The cost for ports is too high.

There is also a cost in sound quality. Voice frames must be decoded and converted to analog (or at least decompressed to PCM if the interface is digital) to enter the PBX. Coming back out, the voice is again compressed and perhaps converted from analog to digital. Every conversion or compression adds distortion and noise, reducing the sound quality. These decrements in quality are important in countries where the carrier is strict about how many "demerits" a voice connection can accumulate and still be connected to the public network.

Voice frame switching, a feature in some VFRADs, allows a call connection to pass through an intermediate VFRAD without terminating there on an analog (or digital) port (Fig. 44). Since the information stays in compressed voice frames, there is no effect on sound quality—a call could pass through many nodes and sound the same. Of course more nodes implies more delay overall, which is a loss in quality and gets its own demerits.

Say a branch office A calls a phone (extension 777 at branch B) that is not colocated with the central VFRAD. When the call connection message from the branch reaches HQ, it is handled similarly to a call request from a local HQ port: the central VFRAD looks up the destination number in a routing table. The second (central) VFRAD makes the same kind of routing decision as the originating VFRAD based on the called phone number.

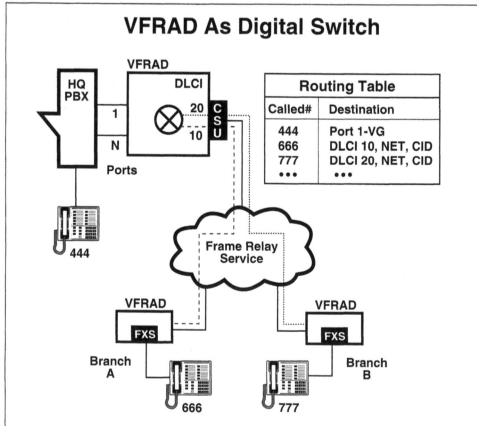

VFRAD As Digital Switch

Routing Table	
Called#	Destination
444	Port 1-VG
666	DLCI 10, NET, CID
777	DLCI 20, NET, CID
•••	•••

Fig. 44. After the VFRADs can interpret dialed digits, they have some of the tools needed to switch calls. Applying the routing more generally, the destination can be a port/DLCI/CID combination leading to another VFRAD as well as a local voice port.

If the call had come from extension 444 at HQ, to Ext. 777, the destination from the routing table would have been the same: DLCI 10 on the Network port.

In this case, the routing is to a VFRAD at another location, via a specific port (the network port) and PVC (DLCI 20). Subframe channel numbers may be negotiated between adjacent VFRADs, using proprietary methods, or assigned at installation. The FRF IA covers only permanently provisioned connections; there is no method in the IA for the FRADs to negotiate any parameters.

The VFRAD at HQ sets up a virtual digital cross-connection between the "calling" and "called" PVCs (DLCIs 10 and 20). One PVC is the source of the voice call request; the second is found from the information in the routing table and may be on a different physical link or even on a different frame relay network. For the duration of the call, VoFR frames received on one side are passed directly to the other, without further processing that would impair voice quality (except the inevitable delay). The virtual connection uses no ports, on the VFRAD or the PBX, and imposes minimum delay.

The VFRAD at HQ acts as a frame relay switch, in a sense, except that it must consider the channel ID as well as the port and DLCI while routing frames. A standard FR switch cannot read the CID.

Most of the configuration is done manually, which can be tedious. The software that performs the routing is not simple, and can be limited as to the number of phones or ports in a network, possible dialing plans, and so on. This feature will be made obsolete by switched virtual circuits.

PSTN Connection

To gain the benefits of making long distance calls over a frame relay network, there must be an attachment to CO lines (Fig. 45). In many locations those interfaces could be on an existing PBX or key system—the VFRAD connects to the PBX as a trunk.

In some locations there may be a desire to connect the VFRAD directly to the local loop from the CO. Those VFRADs will need "full strength" interfaces designed like those on a channel bank (compliant with Pub. 43801, for example). VFRADs designed for short loops or in-house wiring may not qualify for the necessary certifications ('homologation') by carriers or national authorities.

Short-loop designs are very attractive for their low power and low cost. Just don't misapply them to CO connections—keep them on the PBX or a telephone.

Digital DS-1 Interface

Some resources needed in a VFRAD, most notably the voice compression processors, are proportional to the number of channels. The same is true of analog interfaces: each channel requires another physical port and associated components. Digital voice interfaces can offer extra ports for no more hardware or money (Fig. 46).

At about six voice channels the cost of analog ports roughly equals the hardware cost of a digital port. The Digital Signal level 1 (DS-1) is a T-1 or E-1. (See the Author's "Guide to T-1 Networking" for the physical and electrical details of a DS-1.) So, from about 8 voice ports up to 24 or 30, the additional ports are practically free.

Up to 4 voice ports will be cheaper in analog. More than eight voice channels, now or expected in the future, indicates a T-1 or E-1 is the best choice. Equipment from different vendors will vary in their break-even points.

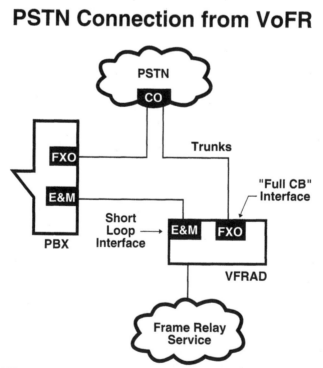

PSTN Connection from VoFR

Fig. 45. VFRADs designed for connection to phones and PBXs in the same building may use a "short loop" interface which cuts size, power, and cost by reducing voltages, lightening protection, etc. Direct connection to central office trunk requires full compliance with standards for channel bank interfaces that drive long local loops.

The DS-1 is always uncompressed, PCM encoded voice: the same stuff as a channel bank. Compression is applied to the signal before it is put into frame relay frames.

Signaling on older PBXs most often will be robbed bit, on T-1, or the similar Channel Associated Signaling on E-1. These bit-oriented signaling methods can use the faster DTMF signaling for dialing, but almost all will accept dial pulses in the form of toggled A/B signaling bits on the digital interface. The A bit will be needed for supervision, to report on/off-hook status.

New equipment is migrating users toward message-oriented signaling, or common channel signaling. A popular format is ISDN primary rate interface (PRI) which uses channel 24 of a T-1 to carry message frames. The standards are based on ISDN, Q.921 to Q.933 and others. They include DSS-1 (Digital Subscriber Signaling, system 1) in the US and Q.sig (a similar set, not quite finalized by ITU at this writing) in Europe. England and France have regional variants, as do some other countries.

All common channel signaling methods are based on HDLC frames, which are easily encapsulated in VoFR subframes. Thus common channel signaling can share the frame relay frames with voice.

VFRADs working according to the VoFR IA simply transfer signaling messages between two switches, treating them like point-to-point data traffic. PBXs of the same make can use all their proprietary phone features over frame relay tie trunks by exchanging proprietary signaling messages.

The procedures defined by the IA do not deal with connecting a call from an analog port at one end to a digital port at the other. Some available equipment can do it, but in a proprietary way.

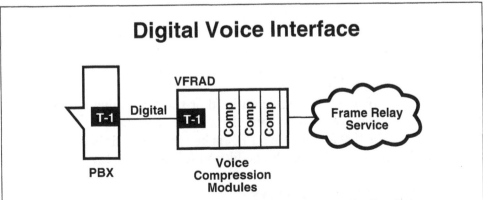

Digital Voice Interface

Fig. 46. A digital voice interface is the standard T-1 or E-1 format compatible with channel banks and robbed bit or channel-associated signaling. Latest PBXs use ISDN, which has a message-oriented signaling format that replaces one voice channel on the digital interface.

Voice Interworking

Despite the incompleteness of the initial VoFR Implementation Agreement, universal interworking is the ultimate goal. The combinations of interfaces at the ends of a PVC could be:

Remote	Central	
FXS	FXO	straight "foreign exchange"
FXS	FXS	automatic and manual ring down
FXS	DS-0	phone to digital PBX trunk
FXS	E&M	phone to analog PBX trunk
E&M	E&M	analog tie trunk
DS-0	DS-0	digital tie trunk
E&M	DS-0	analog to digital tie trunk
FXO	FXO	"Fake" tie trunk on PBX line ports

The goal is to make any interface interoperate with any other. This amounts to a need for a protocol conversion between the channel bank signaling procedures (CAS type) and the procedures toward the frame relay network (per the IA).

To accomplish that conversion there must be a 'state machine' in each VFRAD to control the signaling function and be responsible for translating between the signaling procedures, like E&M and FXS or other pairs. A state machine is a form of software process that remembers what has happened in the past and uses that information to help select future actions from a carefully defined set.

With the control states defined, and the state machines able to translate among the signaling protocols at different interfaces, the frame relay network will support connectivity between any two voice interfaces. That is, an E&M trunk interface on a PBX will interoperate with an ordinary phone (or key system) on an FXS interface.

But please don't ask for music on hold or the "executive barge-in override lock-out" feature offered on some PBX you once managed. The amount of software for the 300 features in a PBX is beyond belief. It is not within the range of what transmission products, like VFRADs, are likely to support.

VoFR Equipment

The voice over frame relay market is moving extremely fast at this time. You'll have to consider this section a blurred snapshot, rather than a detailed buying guide to specific products.

This section is more conceptual, intended to define types of equipment and where they would fit into applications described in the examples above. Don't make more than a little commitment to the classifications. What is described as a branch office device may be perfectly adequate at some regional centers or even headquarters. It depends on your application.

VoFR for Branch Offices

The first VoFR products were small, intended for branch offices and similar sites on at least one end of a circuit. Today, remote VFRADs have 1 to 4 analog voice grade interfaces. These may be FXS with ARD; FXO; or E&M.

The current bandwidth requirement for most branch offices is 56 or 64 kbit/s for frame relay access. This is sufficient for these voice channels and some data traffic.

A LAN port is highly desirable in most branch offices. The choice of Ethernet or token ring will depend mostly on the make of the host computer. While each CPU vendor favors one LAN type, you can find examples of the other on any given computer. Some corporations support both LAN types and multiple routing protocols.

The microprocessor that controls the frame relay interface can perform LAN routing. Support for IP, IPX, and AppleTalk (on Ethernet) are known to be available in small FRADs. The same hardware also supports many different protocols on serial data ports. How many and which protocols depends on software.

Most FRAD vendors offer a choice of software versions. In some cases the additional features are additive; that is, you pay more to get a wider selection—up to the full set of every available feature. Other vendors, perhaps with bulkier code or less memory, offer only a subset of all features in any given software load—to add features you must also drop some.

Most hardware now includes FLASH memory for the operating software. This is the non-volatile storage place for that software you bought. Its advantage is that FLASH can be erased electrically, and loaded with new software from a remote location. Older hardware designs use EPROM, which is erased by exposure to ultraviolet light (not something you can do remotely!). FLASH lets the vendor fix bugs—and sell you more features in the future—without the expense of sending a person to upgrade every site.

Branch offices often have a mix of equipment, from different periods in the company's history, bought for different jobs, from different vendors. Its no surprise they run different protocols. With their cost, and often critical relationship to doing business, these pieces of equipment will be there for many years. And so will their protocols.

In addition to telephones, a bank branch usually contains equipment that needs at least four data communications types:

1. SDLC or Burroughs Poll/Select (or both!) for account transaction processing at the window or counter.

2. Bisync to control and monitor the cash dispenser machine.

3. Async for the alarm system.

4. a LAN, either token ring or Ethernet, for the PCs on the desks of bank officers, and possibly other equipment.

All of these data protocols can be carried on frame relay. The FRAD translates older data formats (SDLC, BSC, X.25, etc.) to the FR format. In addition, it may accept FR data directly, for concentration, and convert between data protocols.

Adding voice interfaces may mean adding a VFRAD, but not replacing the exiting data FRAD. Note that the VFRAD almost always should be installed directly on the frame relay line—there is usually only one. The data FRAD goes behind the VFRAD, so the VFRAD can fragment the long data frames and encapsulate them in VoFR subframes.

An integral CSU makes for a nice, compact package.

Regional Office Needs

The next larger site may be characterized by multiple frame relay access trunks with speeds up to 1 or 2 Mbit/s, 6 or more voice ports, or both. From the data viewpoint, the number of serial ports can be more than 4 but probably less than 16. These models may be offered with T-1 or E1 voice interfaces.

Regional-sized VFRADs are physically larger than remote terminals, to make room for the additional ports and voice processing hardware. Of course the voice compression and signaling methods must be interoperable with the other equipment in the network. Today that means all equipment is from the same vendor.

Internally, the construction is modular to allow for various size implementations, easy intermixing of different interface types, and later addition of enhancements.

Considerations of protocols and FLASH memory are the same as for the branch VFRAD. Internal T-1 CSUs are just starting to be announced.

Central Site Solutions

The earliest packet voice products worked only between small units with a few voice ports. For each remote site that had a voice device, the central site needed to have one of the same device. The headquarters communications room became a jumble of little boxes if the network grew very large.

The same problem of a point-to-point limit on communications once existed in the data world. That problem was solved, in many different ways:
— large TDM multiplexers with multiple trunks;
— large routers, with multiple WAN and LAN ports;
— large packet and frame switches;
— ATM and cell relay switches.

In VoFR, the central site product needs a large number of analog voice ports; up to 60 are available in one unit. This device is based on a shelf with multiple cards inserted front and back. By choice of the modular cards, some of those voice ports can be traded for serial ports when data connectivity is needed.

Voice compression hardware is also modular. Cards are available with anywhere from 4 to 15 DSP chips.

While these large voice nodes are usually placed in the center of star topologies, they may also be connected to each other, either on the FR network or via a leased line. In this configuration they perform voice compression and digital speech interpolation (DSI) to increase the capacity of the line by a factor of 8, 16, or more.

At this writing, the largest voice FRADs are not as sophisticated with data as are the smaller VFRADs. Consequently, practical networks often have multiple devices at the central site, each performing a specialized function. The VFRAD contributes voice compression, data fragmentation, and sometimes frame relay switching.

The network in Fig. 47 consists of terminals and host computers that communicate via standard protocols encapsulated in frame relay. Both were designed for a leased line networks, not frame relay.

In this design, data protocols had to be included in the VFRADs, as well as voice functions, for LAN and serial connections at the remote sites. In order to carry all traffic consistently, and remain compatible with the necessary equipment, some protocol frames were encapsulated twice in frame relay.

Similar network designs have used X.25 as the first encapsulation, and then frame relay. In these cases, the central site needs a large X.25 switch that supports 'Annex G' format for carrying X.25 over frame relay. The block diagram looks very similar.

Note that only frame relay (or frame relay and X.25) and the voice traffic terminate on the VFRADs. If the frame and packet switching is done in another product, this observation leads to a new product category, the transcoder.

Bulk Voice Transcoder

A VoFR transcoder converts between continuous PCM (Pulse Code Modulation) and a serial stream of VoFR frames. Some versions are fixed configurations: one DS-1 for PCM voice, one serial stream for VoFR frames. Other designs are modular, with space for several voice compression modules. Each VCM might handle 6 to 30 voice channels.

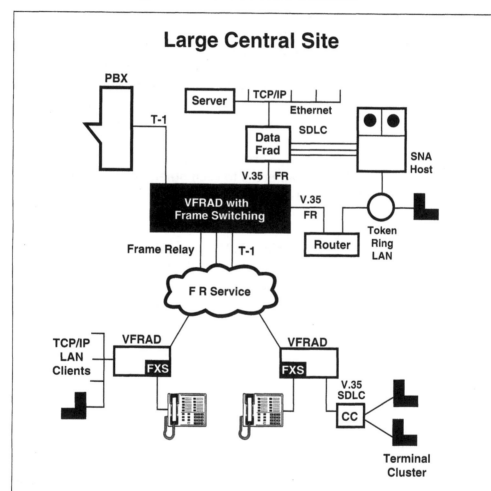

Large Central Site

Fig. 47. The largest VFRADs offer frame relay switching and standard data encapsulations. LAN traffic, in standard format, is fragmented by the remote VFRAD in subchannel. Central VFRAD reassembles full frame, sends it in frame relay (RFC-1490) to a data FRAD or router for media conversion onto a LAN. Legacy data, like SDLC, may be carried in proprietary format and delivered as SDLC, or converted to LLC-2 before encapsulation and delivery on a token ring LAN.

Arriving VoFR frames are decompressed to PCM, but not converted to analog. Eliminating the analog ports reduces the bulk considerably—and the cost. Since the ABCD bits are present on the digital voice port, there is no conversion to be done.

Data isn't terminated on a transcoder. However it might have to pass data frames through in order to fragment them. This function will depend on the application. If there are no slow (56 kbit/s) links in the network, and no very long token ring frames, then fragmentation may not be necessary.

Packet Voice Network Options

Frame relay isn't the only packet medium on offer. Packetized voice has been tried on many types of networks.

X.25

When packetized voice was young, the only packet switching networks were based on X.25 procedures. Error correction performed on every X.25 link introduces large delays. So packet voice servers use TDM channels on private networks. Not trying to get voice over X.25 still seems like a good idea. But with improved compression algorithms, backbone links among X.25 packet switches boosted to 1.5 or 2 Mbit/s, and reduced expectations among users.... Who knows?

IP (on the Internet)

1996 was the year of the Internet phone. Trade publications covered the topic constantly. Software vendors, from the largest on down, offered Application Programming Interfaces (APIs) of all sorts. Hardware and semiconductor vendors announced endless variations and combinations of chips and boards to get your voice calls on the 'Net.

If there were ever too much of a good thing, InterPhone options must be. The only way to be sure a connection will work properly is to have the same hardware and software at both ends.

Then there is the problem (read 'expense') of a full-time Internet access connection in order to receive calls at random times. The Internet generally has no dial-out capabilities so you must be connected first to receive a voice call. If your location hosts a World Wide Web site or on-line Email you probably have full-time access.

Performance is a more serious issue than with frame relay or X.25 services, which are connection-oriented and display a modest amount of transmission delay compared to the Internet.

The Internet is the original connectionless service, based on routers (another form of packet switch). "Connectionless" means possible variations in route from packet to packet, which increases variability in delay or jitter and complicates almost every aspect of voice transmission.

So far, Internet telephones have been popular with experimenters, among friends, within families. The quality is just good enough to be worth the incremental price of hardware and software if you're on the 'Net anyway. It is not a business tool, at this writing, except for voice annotations to documents or voice messages which are not in 'real time.'

Internet phoning may be moving up scale, however. Serious players in dial access and voice processing have announced and started to deliver specialized servers that link standard telephones with the Internet. The process involves steps similar to VoFR—digitization, compression, and signaling— with the addition of dial-out capability. The server, which is always connected to the Internet, uses the PSTN to complete calls.

On the 'Net, signaling is not quite the same as on standard telephone circuits. So the IP/voice server does conversions between dialing tones (typically) and finding an IP address for the called party. This is not straight forward.

Because of the shortage of IP network addresses, there has been almost a rationing (mostly voluntary, some encouraged by pricing) of individual IP addresses. The IP protocol suite has been enhanced to allow individuals to pick up an IP address from the company pool, so to speak, when signing on to the network. At sign-off the address goes back into the pool. Next time it is unlikely that the user will get the same IP address.

IP phone makers have started offering directory services where users may register their current IP address, in real time. That is, part of the automated sign-on process in the workstation is to send the user's name and present IP address to the directory server. At sign-off, another message cancels the listing. This process not only tells others what your address is, it also lets people know if you are available (on-line) or not.

Each address server deals only with one vendor's Internet phone equipment. That is bad because it fragments the user universe. At present, it doesn't matter much because each vendor's equipment is proprietary and won't interoperate with any other vendor's. Standardization is expected, but not without a struggle among the various vendors.

ATM

The ATM Forum has concentrated largely on circuit emulation for voice in TDM channels. It has limited its interest in true packetized voice to one working group related to ATM for trunk lines among cellular phone base stations.

Attempts are being made by the Frame Relay Forum to produce relatively easy interworking between the technologies. The precedent of the two Forums picking different data encapsulation formats is not a positive sign.

Frame Relay

The most likely medium you'll select for packetized voice, after any kind of analysis, is frame relay.

Frame relay is carrying voice successfully, worldwide. It's happening now.

Frame relay has technical advantages in delay, jitter, and frame loss compared to other network types. With modern compression algorithms (standard or proprietary) the voice quality is more than good enough.

And where more appropriate than the bottom line of the book to state it again, voice over frame relay can save you money.

Acronyms

Networking Acronyms

This glossary tries to define acronyms in telecom that are common enough for a network manager to come across, now or in the next generation of equipment. Among the true acronyms listed here in all capitals are some abbreviations which appear, as they normally do, in lower case letters. Numerical items are listed last.

Note that the Index doesn't cover this glossary.

Many items carry a source reference in parentheses. (802.x) = IEEE LAN standard; (Tel) = telephone terminology; (SS7) = signaling system 7; (Layer x) = in OSI model; (A.NNN) = ITU-T recommendation; other references are acronyms, listed here.

A

A	Ampere, unit of electrical current.
AAL	ATM Adaptation Layer, two sublayers concerned with segmenting large PDUs into ATM cells; type 1 = CBR, 2 = VBR, 3/4 = connectionless services, 5 = for LAN and FR frames, 6 = MPEG video. See also SAR, CPCS.
AAP	Alternate Access Provider, carrier other than local telco that can provide local loop into IXC or LEC; CAP.
AAR	Automatic Alternate Routing, failure recovery.
ABAM	Order code for the 22 AWG shielded twisted pair cable used for manual cross connects of DS-1 through DS-2 signals.
ABCD	Signaling bits, for robbed bit signaling with ESF; only A and B are available with SF (Tel).
ABM	Asynchronous Balanced Mode (Layer 2).
ABR	Available Bit Rate, 'best effort' BoD (ATM) with flow control and 'CIR.'

ABS   Alternate Billing Service; credit card, 3rd party, etc. (SS7).

a.c.   Alternating Current, the form of analog voice signals, ringing, and power lines; also 'ac'.

ACARS   Aircraft Communications and Reporting System, VHF network to extend ADNS.

ACCTS   Aviation Coordinating Committee for Telecommunications Services, part of ARINC.

ACD   Automatic Call Distributor, PBX function or machine to spread calls among phones.

ACELP   Algebraic CELP, form of voice compression (G.729, etc.).

ACF   Access Control Field, first byte in ATM header (802.6).

ACF   Access Coordination Function, tariffed service where AT&T obtains local loops between customer premises and AT&T serving office.

ACF   Advanced Communications Function, SNA software.

ACK   Positive Acknowledgment, message or control bytes in a protocol; report a frame was received OK.

ACL   Applications Connectivity Link, Siemens' PHI.

ACM   Address Complete Message, signaling packet equivalent to ring-back tone or answer (SS7).

ACR   Actual (Average) Cell Rate, ATM interface parameter.

ACSE   Association Control Service Element (OSI).

ACT   ACTivate, BRI control bit, turns on NT.

ACT   Applied Computer Telephony, Hewlett-Packard's PHI.

A/D   Analog to Digital, usually a conversion of voice to digital format.

ADCCP   Advanced Data Communications Control Procedure, ANSI counterpart to HDLC.

ADCR   Alternate Destination Call Redirection, service diverts calls to second site (AT&T).

ADCU   Association of Data Communications Users.

AEPA   ATM Endpoint Address, globally unique user address on ATM network.

ADM   Add/Drop Multiplexer, node with 2 aggregates that supports data pass-through.

ADM   Asynchronous Disconnect Mode (Layer 2).

ADNS   ARINC Data Network Service, a packet network.

ADPCM   Adaptive Differential Pulse Code Modulation, a form of voice compression that typically uses 32 kbit/s.

ADSL   Asymmetric Digital Subscriber Line, local loop technology for unequal transmission speeds in CO-->CP and CP-->CO directions (HDSL).

AESA   ATM End System Address.

AF   ATM Forum.

AFI   Authority and Format Identifier, part of network level (NSAP) address header (MAP, ATM).

AFIPS   American Federation of Information Processing Societies.

AFR   Alternate Facility Restriction.

AI   Alarm Indication.

AIB   Alarm Indication Bit, BRI control bit.

AIM   ATM Inverse Multiplexing.

AIN   Advanced Intelligent Network, carrier offering more than 'pipes' to users.

AIS   Alarm Indication Signal, unframed all 1's (Blue Alarm) sent downstream (to user) from fault site. Also used as T-1 keep-alive signal.

ALS   Active Line State, possible status of FDDI optical fiber.

AMI   Alternate Mark Inversion, line coding for T-1 spans where 0 (space) is no voltage and successive 1s (marks) are pulses of opposite polarity. See also DMC, NRZ, 4B/5B.

AMIS   Audio Messaging Interchange Specification, for voice mail.

AMPS   Advanced Mobile Phone Services (System), analog cellular in N.A.

AMS   Audio/visual Multimedia Services (ATM).

AN   Access Node.

ANI   Automatic Number Indication, display of calling number on called phone.

ANM   ANswer Message, signaling packet returned to caller indicating called party is connected (SS7).

AN   Access Network, OSP like local loops, etc.

ANP   AAL-CU Negotiation Procedure, sets up ATM call.

ANR   Automatic Network Routing (APPN).

ANSI   American National Standards Institute, the US member of the ISO.

ANT   Alternate Number Translation, ability to reroute 1-800 calls on NCP failure.

AOS   Alternate Operator Service, non-telco firm responding to "dial zero."

AP   Action point, SDN switch located closest to customer site.

APDU   Application PDU (OSI).

API   Application Programming Interface, software module or commands to separate OS or network from application.

APP   ATM Peer-Peer connection, between adjacent devices; half duplex.

APPC   Advanced Program to Program (Peer-to-Peer) Communications, session level programming interface (APPN).

APPI   Advanced Peer to Peer Internetworking, cisco version of APPN that encapsulates in IP.

APPN   Advanced Peer-to-Peer Networking, IBM networking architecture.

APS   Automatic Protection Switch.

AR   Access Rate, speed of a channel into a backbone network.

ARAP   AppleTalk Remote Access Protocol.

ARD   Automatic Ring Down, lifting your phone rings the fixed far end without dialing.

ARINC   Aeronautical Radio, Inc., operator of private airline networks.

ARP   Address Resolution Protocol, a way for routers to translate between different forms of protocol addresses or domains.

ARPA   Advanced Research Projects Agency, created Arpanet packet network, first X.25 net; folded into NSFnet in 1990.

ARQ   Automatic Repeat reQuest, for retransmission; an error correction scheme for data links, used with a CRC.

ASAI   Adjunct Switch Applications Interface, AT&T's PHI.

ASCII   American Standard Code for Information Interchange, based on 7 bits plus parity.

ASDS   Accunet Spectrum of Digital Services, AT&T fractional T-1.

ASE   Applications Service Element, protocol at upper layer 7 (SS7, OSI).

ASIC   Application Specific Integrated Circuit, custom chip.

ASN.1   Abstract Syntax Notation #1, language to manage network elements.

ASR   Automatic Send/Receive, a printer with keyboard or a Teletype machine.

ASTN   Alternate Signaling Transfer Network, a CCS6 that backs up CCS7.

ATAP   All Things to All People, the mythical perfect product.

ATD   Asynchronous Time Division, ETSI proposal for pure cell relay, without SONET or other framing.

ATM   Asynchronous Transfer Mode, a type of cell transfer protocol or packet framing.

ATMF   ATM Forum; also AF.

ATMM   ATM Management entity, function in ATM device.

ATN   Aeronautical Telecommunication Network, as in ATN Protocol Architecture used by ARINC.

AU   Administrative Unit, payload plus pointers (SDH).

AUU   ATM User-User connection, end to end.

AUG   AU Group, one or more AUs to fill an STM (SDH).

AUI   Attachment Unit (Universal) Interface, standard connector between MAU and PLS (802.x Ethernet).

Autovon Automatic Voice network, a U.S. military net.

AVS   Available Seconds, when BER of line has been less than 10-3 for 10 consecutive seconds until UAS start.

AWG   American Wire Gauge, conventional designator of wire size.

B

B   Bearer channel, a DS-0 for user traffic (ISDN).

B   Beginning, bit in header of VoFR subframe, for data fragmentation.

B1   SOH byte carrying BIP-8 parity check (SONET).

B2   LOH byte carrying BIP parity check.

B3   POH byte carrying BIP parity check.

B3ZS   Binary 3-Zero Substitution, line coding for DS-3 signal substitutes a 'coding violation' for 4 consecutive zeros.

B8ZS   Binary 8-Zero Suppression, substitutes 000+-0-+ for 00000000 to maintain ones density on T-1 line.

BAN   Boundry Access Node, edge device in APPN.

BAsize   Buffer Allocation; number of octets in L3-PDU from DA to Info, +CRC if present (SMDS, ATM).

BATT   Battery, the -48 (-40 to -52) V d.c. supply in the CO

Bc   Committed Burst, amount of data allowed in time T=Bc/CIR without being marked DE.

BCC   Bellcore Client Company, one of 7 RBOCs who owned Bellcore until 1997.

BCC   Block Check Code, a CRC or similarly calculated number to find transmission errors.

BCD   Binary Coded Decimal, 4-bit expression for 0 (0000) to 9 (1001).

BCM   Bit Compression Mux, same as M44 for ADPCM.

BCN   Beacon, frames sent downstream by station on token ring when upstream input is lost (802.5).

B-DCS   Broadband Digital Cross-connect System, DACS OC-1, STS-1, DS-3 and higher rates only (see W-DCS).

BDLC   Burroughs Data Link Control, layer 2 in Burroughs Network Architecture.

Be   Excess Burst, transient capacity above CIR and Bc in FR net.

BECN   Backward Explicit Congestion Notification, signaling bit in frame relay header.

BER   Bit Error Ratio (Rate), errored bits over total bits; should be <10^7 for transmission lines.

BERT   Bit Error Rate Test(er).

BES   Bursty Errored Second, from 1 to 319 CRC errors in ESF framing, ESB.

BEtag   Beginning/End tag; same sequence number put at head and tail of L3-PDU.

BGF   Basic Global Functions, requirements for ISDN.

BIB   Backward Indicator Bit, field in SUs (SS7).

B-ICI   Broadband ICI, interface between public ATM networks.

BIOS   Basic Input/Output System, part of OS; may include communications functions.

BIP-x   Bit Interleaved Parity, error checking method where each of x bits is parity of every xth bit in data block (x=8 in SONET, 16 in ATM).

B-ISDN   Broadband ISDN, generally ATM access at more than 100 Mbit/s.

B-ISSI   Broadband Inter-Switching System Interface, e.g., between ATM nodes.

BISYNC   Binary Synchronous communications, a protocol; also BSC.

bit   One eighth of something, e.g., half the $ cost of 'shave & a haircut.'

BITS   Building Integrated Timing Supply, stratum 1 clock in CO.

BIU   Basic Information Unit, up to 256 bytes of user data (RU) with RH and TH headers(SNA).

BLERT   BLock Error Rate Test.

BLU   Basic Link Unit, data link level frame (SNA).

BMS   Bandwidth Management Service, AT&T offering like a private network with equipment in CO.

BN   Bridge Number, device identifier in LAN for routing/bridging.

BNA   Burroughs Network Architecture, comparable to SNA.

BNS   Broadband Network Switch, usually ATM or packet based, DS-3 and faster.

BOC   Bell Operating Company, a local telephone company.

BoD   Bandwidth on Demand, dynamic allocation of line capacity to active users.

BOM   Beginning of Message, type of segment (cell) that starts a new MAC frame, before COM and EOM (SMDS).

BOM   Bill Of Materials, list of all parts in an assembly.

BONDING   Bandwidth On Demand INteroperability Group, makers of inverse muxes and standard they adopted.

BOOTP   Boot Protocol, lets station get IP address from server using UDP/IP; enhanced RARP.

BORSHT   Battery feed, Over-voltage protection, Ringing, Signaling (Supervision), Hybrid, Test; classic functions of analog interface.

bps   Bits per second, serial digital stream data rate, now bit/s.

BP/S   Burroughs Poll/Select, legacy data protocol.

BPV   Bipolar Violation, two pulses of the same polarity in a row.

BR   Bureau of Radiocommunications, part of ITU that allocates international spectrum.

BRA   Basic Rate Access, ISDN 2B+D loop.

BRI   Basic Rate Interface, 2B+D on one local loop.

BRT   Broadband Remote Terminal, node with DS-3, OC-1, or faster access into ATM, etc.

BSC   Binary Synchronous Communications, a half-duplex legacy data protocol.

BSN   Backward Sequence Number, sequence number of packet (SU) expected next (SS7).

BSP   Bell System Practice, document associated with equipment used by telcos for I&M, etc.

BSRF   Basic System Reference Frequency, formerly Bell SRF; Stratum 1 clock source of 8 kHz.

BSS   Broadband Switching System, cell-based CO switch for B-ISDN.

BT   British Telecom, primary phone company in the United Kingdom.

BTag   Beginning Tag, field in header of frame whose value should match ETag.

BTAM   Basic Telecommunications Access Method, older IBM mainframe comm software.

BTU   Basic Transmission Unit, LU data frame of RU with RH (SNA).

BUS   Broadcast and Unknown Server, last resort to map MAC to ATM address.

BWB   Bandwidth Balancing, method to reduce a station's access to a transmission bus, to improve fairness (802.6).

C

C   Capacitance, the property (measured in farads) of a device (capacitor) that holds an electrical charge.

C-Plane Control Plane, out of band signaling system for U-Plane.

CA*net   Canadian Academic Network.

CAC   Connection (Call) Admission Control, process to limit new calls to preserve QoS (ATM).

CAD   Computer Aided Design, drafting on computers.

CAE   Computer Aided Engineering.

CALM   Connection Associated Layer Management, F5 flows (ATM),

CAM   Computer Aided Manufacturing.

CAP   Carrierless Amplitude and Phase modulation, a modem technique applied at up to to 50 Mbit/s in LANs and HDSL.

CAP   Competitive Access Provider, alternative to LEC for local loop to IXC or for dial tone.

CAS   Channel Associated Signaling, bits like ABCD tied to specific voice channel by TDM.

CASE   Common Application Service Elements, application protocol (MAP).

CAT   Category, often with a number (CAT-3) to indicate grade, as of UTP wiring.

CATV   Community Antenna Television, cable TV.

CB   Channel Bank, 24-port voice multiplexer, to T-1 interface.

CBEMA   Computer and Business Equipment Manufacturers Association.

CBR   Constant (Continuous) Bit Rate, channel or service in ATM network for PCM voice or sync data in a steady flow with low variation in cell delay; emulates TDM channel.

CBX   Computerized Branch eXchange, PABX.

CC   Call Control.

CC   Continuity check, OA&M cell (ATM).

CC   Cluster Controller, for group of dumb terminals (SNA).

CCBS   Completion of Calls to Busy Subscribers, supplementary service defined for ISDN.

cch   Connections per Circuit-hour, in Hundreds.

CCIS   Common Channel Inter-office Signaling.

CCITT   Comite Consultatif Internationale de Telegraphique et Telephonique, The International Telegraph and Telephone Consultative Committee, part of ITU that was merged with CCIR in 1993 to form TSS.

CCIR   International Radio Consultative Committee, sister group to CCITT, became part of TSS (1993).

CCR   Commitment, Concurrency, and Recovery (OSI).

CCR   Current Cell Rate (ATM).

CCR   Customer Controlled Reconfiguration, of T-1 lines via DACS switching.

CCS   Common Channel Signaling.

CCS   Common Communication Subsystem, level 7 applications services (SNA).

CCSA   Common Control Switching Arrangement.

CCSS   CCS System, usually with a number.

CCS6   CCS system 6, first out of band signaling system in N.A. (CCIS).

CD   Carrier Detect, digital output from modem when it receives analog modem signal on phone line.

CD   Compact Disk, as in CD-ROM data storage.

CD   Count Down, a counter that holds the number of cells queued ahead of the local message segment (802.6).

CDMA   Code Division Multiple Access, spread spectrum; broadcast frequency changes rapidly in pattern known to receiver.

CDPD   Cellular Digital Packet Data.

CDT   Cell Delay Tolerance, ATM parameter.

CDV   Cell Delay Variation, ATM UNI traffic parameter.

CE   Circuit Emulation, ATM function to carry TDM circuits (T-1, etc.) on cell stream.

CE   Connection Element (ATM, LAP-D).

CEI   Connection Endpoint Identifier (ATM, LAP-D).

CELP   Code-Excited Linear Predictive coding, a voice compression algorithm used at 16 kbit/s, 8 kbit/s, and slower rates.

CEN   Committee for European Standardization.

CENELEC   Committee for European Electrotechnical Standardization.

CEP   Connection End Point (ATM).

CEPT	Conference on European Posts & Telecommunications (Conference of European Postal and Telecommunications administrations), a body that formerly set policy for services and interfaces in 26 countries.
CERN	Nuclear research facility (particle physics) in Geneva, Switzerland.
CES	Circuit Emulation Service, carries TDM channel on cells (ATM).
CES	Connection Endpoint Suffix, number added by TE to SAPI to make address for connection; mapped to TEI by L2.
CFA	Carrier Failure Alarm, detection of red (local) or yellow (remote) alarm on T-1.
CGA	Carrier Group Alarm, trunk conditioning applied during CFA.
CHAP	Challenge Handshake Authentication Protocol, log-in security procedure for dial-in access.
CI	Congestion Indication.
CI	Connection Identifier, frame or cell address.
CI	Continuation Indicator (ATM).
CI	Customer Installation, all the phone equipment and wiring attached to the PSTN; see CPE.
CIB	CRC Indication Bit, 1 if the CRC is present, 0 if it is not used (SMDS).
CICS	Customer Information Control System, IBM mainframe comm software with data base.
CID	Channel Identifier, subframe address (VoFR).
CIDR	Classless Inter-Domain Routing (IP).
CIF	Cells In Flight, number of cells sent before first cell reaches far end.
CIF	Cell Information Field, 48 byte payload in each cell (ATM).
CIR	Committed Information Rate, minimum throughput guaranteed by FR carrier.
CIT	Computer Integrated Telephony, DEC's PHI.
CL	Common Language, Bellcore codes to identify equipment, locations, etc.
CL	ConnectionLess.CLASS Custom(er, ized) Local Area Signaling Services; ANI, call waiting, call forwarding, trace, etc.
CLEI	Common Language Equipment Identifier, unique code assigned by Bellcore for label on each CO device.
CLID	Calling Line IDentification, ANI.
CLIP	Calling Line Identity Presentation, ISDN-UP service to support ANI.
CLIR	Calling Line Identity Restriction, feature where caller prevents ANI (ISDN-UP).
CLIST	Command List, similar to .BAT file (SNA).
CLLM	Consolidated Link Layer Management (802).
CLN	ConnectionLess Network, packet address is unique and network routes all traffic on any path(s); e.g., most LANs like IP.
CLNAP	CLN Access Protocol.
CLNIP	CLN Interface Protocol.
CLNP	ConnectionLess mode Network (layer) Protocol; see CLN.
CLNS	ConnectionLess mode Network (layer) Service, ULP (SNA).
CLP	Cell Loss Priority, signaling bit in ATM cell (1=low).
CLR	Cell Loss Ratio (ATM).
CLS	Connectionless Service.

CLTS	ConnectionLess Transport Service, OSI datagram protocol.
CMA	Communications Managers Association.
CMD	Circuit Mode Data, ISDN call type.
CMDR	Command Reject, similar to FRMR (HDLC).
CMI	Coded Mark Inversion, line signal for STS-3.
CMI	Constant Mark, Inverted; line coding for T-1 local loop in Japan.
CMIP	Common (network) Management Information Protocol, part of the OSI network management scheme, connection oriented.
CMIS	Common (network) Management Information Service, runs on CMIP (OSI).
CMISE	CMIS Element.
CMOL	CMIP Over LLC, reduced NMS protocol stack.
CMOS	Complementary Metal Oxide Semiconductor, low power method (lower than NMOS) to make ICs.
CMOT	CMIP over TCP/IP.
CMT	Connection Management, part of SMT that establishes physical link between adjacent stations (FDDI).
CND	Calling Number Delivery, another name for CLID, one of the CLASS services.
CNF	Confirmed (OSI).
CNIS	Calling Number Identification Service, provide, screen, or deliver CPN or caller ID (ISDN).
CNLP	Connectionless Protocol.
CNLS	Connectionless Service.
CNM	Communications Network Management (SNA).
CNM	Customer Network Management (Bellcore).
CO	Central Office, of a phone company, where the switch is located; the other end of the local loop opposite CP.
C-O	Connection Oriented.
COC	Central Office Connection, separately tariffed part of circuit within a CO.
COCF	Connection Oriented Convergence Function, MAC-layer entity.
CODEC	COder-DECoder, converts analog voice to digital, and back.
COFA	Change of Frame Alignment, movement of SPE within STS frame.
CO-LAN	Central Office Local Area Network, a data switching service based on a data PBX in a carrier's CO.
.com	Commercial, first Internet address domain of businesses.
COM	Continuation Of Message, type of segment between BOM and EOM (ATM, SMDS).
comm	Communications.
CON	Connection-Oriented Network, defines one path per logical connection (FR, etc.).
CONP	Connection mode Network layer Protocol.
CONS	Connection-Oriented Network Services, ULP (SNA).
COS	Class Of Service.
COS	Corporation for Open Systems, R&D consortium to promote OSI in the US; see SPAG.
COSINE	Cooperation for Open Systems Interconnection Networking in Europe.

COT	Central Office Terminal, equipment at CO end of multiplexed digital local loop or line.
COT	Customer-Originated Trace, sends CPN to telco or police (CLASS).
CP	Central Processor, CPU that runs network under center-weighted control.
CP	Control Point, function in APPN node for routing, configuration, directory services.
CP	Customer Premises, as opposed to CO.
CPAAL	Common Part AAL, may be followed by a number to indicate type.
CPCS	Common Protocol Convergence Sublayer, pads PDU to N x 48 bytes, maps control bits, adds FCS in preparation for SAR.
CPE	Customer Premises Equipment, hardware in user's office.
CPI	Computer-PBX Interface, a data interface between NTI and DEC.
CPIC	Common Programming Interface for Communications, a software tool for using LU6.2 adopted by X/Open as a standard.
CPN	Calling Party Number, DN of source of call (ISDN).
CPN	Customer Premises Node (or Network), CPE
CPNI	Customer Proprietary Network Information, customer data held by telcos.
CPSS	Control Packet Switching System, subnetwork of supervisory channels (Newbridge).
CPU	Central Processor Unit, the computer.
CR	Carriage Return, often combined with a line feed when sending to a printer.
CRC	Cyclic Redundancy Check, an error detection scheme, for ARQ or frame/cell discard.
CRF	Connection Related Function (ISDN).
CRIS	Customer Records Information System, telco OSS.
CRT	Cathode Ray Tube, simple computer terminal.
CRV	Coding Rule Violation, unique bit signal for F bit in frame 1 of CMI.
CS	Circuit Switched, uses TDM rather than packets.
CS	Convergence Sublayer, where header and trailer are added before segmentation (ATM).
CSA	Carrier Service Area, defined by a local loop length (<12,000 ft) from CO, or from remote switch unit or RT.
CSA	Callpath Services Architecture, for PBX to IBM host interface.
CS-ACELP	Conjugate Structure-ACELP, specifically G.729.
CSC	Circuit-Switched Channel (Connection).
CSDC	Circuit Switched Digital Capability, AT&T version of Sw56. 61330
CSMA	Carrier Sense Multiple Access, a LAN transport method, usually with "/CD" for collision detection or "/CA" collision avoidance; LAN protocols at physical layer.
CSO	Cold Start Only, BRI control bit.
CSPDN	Circuit Switched Public Data Network.
CS-PDU	Convergence Layer PDU, info plus new header and trailer to make packet that is segmented into cells or SUs.
CSTA	Computer Supported Telephony Application, PHI from ECMA.
CSU	Channel Service Unit, the interface to the T-1 line that terminates the local loop.
CT2	Cordless Telephone, second version; digital wireless telephone service defined by ETSI.

CTR	Common Technical Requirements, European standards.
CTS	Clear To Send, lead on interface indicating DCE is ready to receive data.
CU	Channel Unit, plug in module for channel bank.
CU	Composit User, form of AAL for multimedia traffic (ATM).
CUG	Closed User Group.
CV	Coding Violation, transmission error in SONET section.
CVSD	Continuously Variable Slope Delta modulation, a voice encoding technique offering variable compression.
CWC	Center-Weighted Control, A central processor runs a network-wide functions while nodes do local tasks.

D

D	Delta (or Data) channel, 16 kbit/s in BRI, 64 kbit/s in PRI, used for signaling (and perhaps some packet data).
D3	Third generation channel bank, 24 channels on one T-1.
D4	Fourth generation digital channel bank, up to 48 voice channels on two T-1's or one T-1C.
D5	Fifth generation channel bank with ESF.
DA	Destination Address, field in frame header (802).
D/A	Digital to Analog, decoding of voice signal.
D/A	Drop and Add, similar to drop and insert.
DACS	Digital Access and Cross-connect System, a digital switching device for routing T-1 lines, and DS-0 portions of lines, among multiple T-1 ports.
DAMA	Demand Assigned Multiple Access, multiplexing technique to share satellite channels (Vsat).
DARA	Dynamic Alternate Routing Algorithm.
DARPA	Defense ARPA, formerly just ARPA.
DAS	Dual-Attached (Access) Station, device on main dual FO rings, 4 fibers (FDDI).
DASD	Direct Access Storage Device (SNA).
DASS	Digital Access Signaling System, protocol for ISDN D channel in U.K.
DAVIC	Digital Audio Video Interoperability Council.
dB	Decibel, 1/10 of a bel; 10 log (x/y) where x/y is a ratio of like quantities (power).
dBm	Power level referenced to 1 mW at 1004 Hz into 600 ohms impedance.
dBm0	Power that would be at zero TLP reference level, = measurement - (TLP at that point).
dBrn	Power level relative to noise, dBm + 90.
dBrnC	dBrn through a C-weighted audio filter (matches ear's response).
DB-25	25-pin connector specified for RS-232 I/F.
d.c.	Direct Current, used for some signaling forms; type of power in CO.
DCA	Distributed Communications Architecture, networking scheme of Sperry Univac.
DCC	Data Communications Channel, overhead connection in D bytes for SONET management.
DCC	Digital Cross Connect, generic DACS.

DCC	Direct Connect Card, data interface module on a T-1 bandwidth manager.
DCE	Data Circuit-terminating Equipment, see next DCE.
DCE	Data Communications Equipment, 'gender' of interface on modem or CSU; see DTE.
DCS	Digital Cross-connect System, DACS.
DDCMP	Digital Data Communications Message Protocol.
DDD	Direct Distance Dialing, refers to PSTN.
DDS	Digital Data System, network that supports DATAPHONE Digital Service.
DDSD	Delay Dial Start Dial, a start-stop protocol for dialing into a CO switch.
DE	Discard Eligibility, bit in FR header denoting lower priority; as when exceeding CIR or Bc.
DEA	DEActivate, BRI control bit.
DECmcc	Digital Equipment Corp. Management Control Center, umbrella network management system.
DEO	Digital End Office, class 5 CO or serving office.
DES	Data Encryption Standard, moderately difficult to break.
DFC	Data Flow Control, layer 5 of SNA.
DFN	Deutsche Forschungsnetz Verein, German Research Network Association.
DGM	Degraded Minute, time when BER is between 10^{-6} and 10^{-3}.
DHCP	Dynamic Host Configuration Protocol, enhanced BOOTP, negotiates several parameters.
D/I	Drop and Insert, a mux function or type.
DID	Direct Inward Dial, CO directs call to specific extension on PBX, usually via DNIS.
DIP	Dual In-line Package, for chips and switches.
DIS	Draft International Standard, preliminary form of OSI standard.
DISA	Direct Inward System Access, PBX feature that allows outside caller to use all features, like calling out again.
DISC	Disconnect, command frame sent between LLC entities (Layer 2).
DL	Data Link.
DLC	Data Link Connection, one logical bit stream in LAPD (Layer 2).
DLC	Data Link Control, level 2 control of trunk to adjacent node (SNA).
DLC	Digital Loop Carrier, mux system to gather analog loops and carry them to CO.
DLCI	Data Link Connection Identifier, address in a frame (I.122); LAPD address consisting of SAPI and TEI.
DLE	Data Link Escape, ESC.
DLL	Data Link Layer, layer 2 (OSI).
DLS	Data Link Switching, IBM way to carry Netbios and spoofed SDLC over TCP/IP.
DM	Disconnected Mode, LLC frame to reject a connection request (Layer 2).
DMC	Differential Manchester Code, pulse pattern that puts transition at center of each bit time for clocking, transition [none] at start of period for 0 [1] (802.5).
DME	Distributed Management Environment, OSF's network management architecture.
DMI	Digitally Multiplexed Interface, AT&T interface for 23 64 kbit/s channels and a 24th for signaling; precursor to PRI.
DMPDU	Derived MAC Protocol Data Unit, a 44- octet segment of upper layer packet plus cell header/trailer (802.6); see L2PDU.

DMS   Digital Multiplex System.

DMT   Discrete MultiTone, form of signal encoding for ADSL.

DN   Directory Number, network address used to reach called party (POTS, ISDN).

DNA   Digital Network Architecture, DEC's networking scheme.

DNIC   Data Network Identification Code, assigned like an area code to public data networks.

DNIC   Data Network Interface Circuit, 2B+D ISDN U interface, from before 2B1Q.

DNIS   Dialed Number Identification Service, where carrier delivers number of called extension after PBX acknowledges call.

DNR   DCE Not Ready, signaling bit in CMI.

DoD   Dept. of Defense.

DOD   Direct Outward Dialing.

DoV   Data over Voice, modems combine voice and data on one twisted pair.

DP   Draft Proposal, of an ISO standard.

DP   Dial Pulse, rotary dialing rather than DTMF.

DPBX   Data PBX, a switch under control of end users at terminals.

DPC   Destination signal transfer Point Code, level 3 address in SU of STP (SS7).

DPCM   Differential Pulse Code Modulation, voice compression algorithm used in ADPCM.

DPGS   Digital Pair Gain System, multiplexer and line driver to put 2+ channels over 1 or 2 pair of wires.

DPNSS   Digital Private Network Signaling System, PBX interface for common channel signaling.

DPO   Dial Pulse Originate, a form of channel bank plug-in that accepts dial pulses.

DPT   Dial Pulse Terminate, a channel bank plug that outputs pulses.

DQDB   Distributed Queue Dual Bus, an IEEE 802.6 protocol to access MAN's, typically at T-1, T-3, or faster.

DS-0   Digital Signal level 0, 64,000 bit/s, the worldwide standard speed for PCM digitized voice channels.

DS-0A   Digital Signal level 0 with a single rate adapted channel.

DS-0B   Digital Signal level 0 with multiple channels sub-rate multiplexed in DDS format.

DS-1   Digital Signal level 1, 1.544 Mbit/s in North America, 2.048 Mbit/s in CCITT countries.

DS-1A   Proposed designation for 2.048 Mbit/s in North America.

DS-1C   Two T-1's, used mostly by Telcos internally.

DS-2   Four T-1's, little used in US, common in Japan.

DS-3   Digital Signal level 3, 44.736 Mbit/s, carrying 28 T-1's.

DSAP   Destination Service Access Point, address field in header of LLC frame to identify a user within a station address (Layer 2).

DSG   Default Slot Generator, the function in a station that marks time slots on the bus (802.6).

DSI   Digital Speech Interpolation, a voice compression technique that relies on the statistics of voice traffic over many channels.

DSL   Digital Subscriber Line, any of various ways to send fast data on copper loops (ISDN, HDSL, xDSL, etc).

DSP   Digital Signal Processor, specialized chip optimized for fast numerical computations.

VOICE OVER FRAME RELAY   **137**

DSP	Display System Protocol, protocol for faster bisync traffic over packet nets.
DSR	Data Set Ready, signal indicating DCE and line ready to receive data.
DSS1	Digital Subscriber Signaling system 1, access protocol for switched connection signaling from NT to ISDN switch (Q.931 & ANSI T1S1/90-214).
DSU	Digital (Data) Service Unit, converts RS-232 or other terminal interface to line coding for local loop transmission.
DSX-1	Digital Signal cross connect, level 1; part of the DS-1 specification, T-1 or E-1.
DSX-z	Digital Signal cross connect where 'z' may be 0A, 0B, 1, 1C, 2, 3, etc. to indicate the level.
DT	Data Transfer, type of TPDU in ISDN.
DTAU	Digital Test Access Unit, CO equipment on T-1 line.
DTE	Data Terminal Equipment, 'gender' of interface on terminal or PC; see DCE.
DTI	Digital Trunk Interface, T-1 port on Northern Telecom PBX.
DTMF	Dual Tone Multi-Frequency, TOUCHTONE dialing, as opposed to DP.
DTP	Data Transfer Protocol.
DTR	Data Terminal Ready, signal that terminal is ready to receive data from DCE.
DTU	Data Termination Unit, Newbridge TA for data.
DVD	Digital Versatile Disk, CD format for sound, video, and data.
DVI	Digital Video Interactive, applications with large, bursty bandwidth.
DWMT	Discrete Wavelet MultiTone (DMT).
DX	Duplex, a 2-way audio channel bank plug without signaling.
DXC	Digital Cross Connect, DACS.
DXI	Data Exchange Interface, serial protocol for SNMP for any speed.

E

E	'Ear' lead on VG switch, receives signaling.
E	Ending, indicator bit in header of subframe carrying frame fragment.
E-1	European digital signal level 1, 2.048 Mbit/s.
E-ADPCM	Embedded ADPCM, packetized voice with "core" and "enhancement" portion to each frame.
EA	Extended Address, or address extension bit, =1 in last byte of frame header.
EASlnet	Network for European Academic Supercomputer Initiative.
E&M	Earth and Magneto, signaling leads on a voice tie line, also known as Ear and Mouth.
EBCDIC	Extended Binary Coded Decimal Interchange Code, extended character set on IBM hosts.
EC	Echo Canceller.
EC	Error Correction, process to check packets for errors and send again if needed.
EC	Enterprise Controller, new terminal cluster controller, e.g. 3174.
EC	European Community, covered by common telecom standards.
ECC	Error Checking Code, 2 bytes (usually) in frame or packet derived from data to let receiver test for transmission errors.

ECHO   European Clearing House Organization, foreign exchange settlement network, run by SWIFT.

ECL   Emitter Coupled Logic, transistor circuit type optimized for high speed.

ECMA   European Computer Manufacturers Association.

ECN   Explicit Congestion Notification, network warns terminals of congestion by setting bits in frame header (I.122).

ECO   Engineering Change Order, document from designer ordering change in product.

ECTRA   European Committee on (Telecom) Regulatory Affairs, created out of CEPT in 1990 to be regulatory half, as opposed to operational part, of carriers and PTTs.

ECSA   Exchange Carrier Standards Association.

ED   Ending Delimiter, unique symbol to mark end of LAN frame (TT in FDDI, HDLC flag, etc.).

EDI   Electronic Document (Data) Interchange, transfer of business information (P.O., invoice, etc.) in defined formats.

EDSX   Electronic DSX, usually followed by a "-N" for signal level.

EETDN   End to End Transit Delay Negotiation, part of call setup via X.25 (ISDN).

EFCI   Explicit Forward Congestion Indication, flow control method in ATM networks.

EFCN   Explicit Forward Congestion Notification (ATM). See also, FECN.

EFI   Errored Frame Indicator (ATM on fiber channel).

EFT   Electronic Funds Transfer.

EGP   Exterior Gateway Protocol, used between an autonomous user network and the Internet.

EIA   Electronic Industries Association, publisher of standards (e.g. RS-232).

EISA   Extended ISA, 32-bit PC bus compatible with AT style PCs.

EKTS   Electronic Key Telephone Service, ISDN terminal mode.

ELAN   Emulated LAN, logical net based on LANE (ATM).

E&M   Earth and Magneto, signaling leads on a voice tie line, also known as Ear and Mouth.

EMA   Enterprise Management Architecture, DEC's umbrella network management system.

EMI   ElectroMagnetic Interference.

EMS   Element Management System, usually a vendor-specific NMS for a hardware domain (OSI).

EN   End Node, limited capability access device (APPN).

ENQ   Enquiry, control byte that requests a repeat transmission or control of line.

ENTELEC Energy Telecommunications and Electrical Association.

EOC   Embedded Operation Channel, D bytes devoted to alarms, supervision, and provisioning (SONET); control field in M channel of BRI.

EOM   End of Message, cell type carrying last segment of frame.

EOT   End Of Transmission, control byte; preceded by DLE indicates switched station going on hook.

EPSCS   Enhanced Private Switched Communications Service.

ERL   Echo Return Loss.

ERLE   ERL Enhancement, reduction in echo level produced by echo canceller.

ERS   Errored Second, a 1 sec. interval containing 1 or more transmission errors.

ESB   Errored Second type B, new name for bursty ES.

ESC	Escape, an ASCII character.
ESD	ElectroStatic Discharge, electrical "shock" from person or other source that can destroy semiconductors.
ESF	Extended Super Frame, formerly called Fe.
ESP	Enhanced Service Provider, a firm that delivers itsproduct over the phone.
ESS	Electronic Switching System, a CO switch.
ET	Exchange Termination, standards talk for ISDN switching equipment in CO.
ETag	End Tag, field in trailer of frame whose value should match that in BTag.
ETC	End of Transmission Block, control byte in BSC.
ETN	Electronic Tandem Network.
ETO	Equalized Transmit Only, voice interface with compensation to correct for frequency response of the line.
ETR	ETSI Technical Report.
ETS	Electronic Tandem Switching.
ETS	European Telecommunications Standard, published by ETSI.
ETSI	European Telecommunications Standards Institute, coordinates telecommunications policies.
ETX	End of Text, control byte.

F

F	Farad, electrical unit of capacitance.
F	Final, control bit in frame header (Layer 2).
F	Framing, bit position in TDM frame where known pattern repeats.
F1	Flow (of OA&M cells) at level 1, over a SONET section (ATM).
F2	Flow of OA&M cells over a line.
F3	Flow of OA&M cells between PTEs.
F4	Flow of OA&M cells for metasignaling and VP management.
F5	Flow of OA&M cells specific to a logical connection on one VPI/VCI.
FACS	Facility Assignment Control System, for telco to manage outside plant (local loops).
FAD	Factory Authorized Dealer.
FADU	File Access Data Unit (OSI).
FAS	Facility Associated Signaling, D channel is on same interface as controlled B channels (ISDN).
FAS	Frame Alignment Signal, bit or byte used by receiver to locate TDM channels.
FASTAR	Fast Automatic Restoration, of DS-3s via DACS switching (AT&T).
FAX	Facsimile.
FB	Framing Bit.
FC	Frame Control, field to define type of frame (FDDI).
FCC	Federal Communication Commission, regulates communications in US; also FastComm Communication Corp.

FCOT   Fiber optic Central Office Terminal.

FCS   Frame Check Sequence, error checking code like CRC (Layer 2).

FDDI   Fiber Distributed Data Interface, 100 Mbit/s FO standard for a LAN or MAN.

FDL   Facility Data Link, part of the ESF framing bits available for user data,
             in some cases.

FDM   Frequency Division Multiplexer.

FDMA   Frequency Division Multiple Access, wireless access where each channel has separate
             radio carrier frequency.

FDX   Full DupleX, simultaneous transmission in both directions.

Fe   Extended framing ("F sub e"), old name for ESF.

FEA   Far End Alarm, repeating bit C-3 in DS-3 format
             identifies alarm or status.

FEBE   Far End Block Error, alarm signal (count of BIP errors received; ATM uses Z2 byte in
             SONET LOH).

FEC   Forward Error Correction, using redundancy in a signal to allow the receiver to correct
             transmission errors.

FECN   Forward Explicit Congestion Notification, signaling bit in frame relay header.

FEP   Front End Processor, peripheral computer to mainframe CPU, handles
             communications.

FERF   Far End Receive Failure, alarm signal (ATM).

FEXT   Far End Cross Talk.

FGA   Feature Group A, set of signaling and other functions at LEC-IXC interface; also defined
             for FGB to FGD.

FH   Frame Handler, term in standards for FRS or network.

FIB   Forward Indicator Bit, field in SUs (SS7).

FID   Format Identification, bit C-1 in DS-3 format shows if M13, M28, or
             Syntran signal.

FIFO   First In First Out, buffer type that delays bit stream.

FIPS   Federal Information Processing Standards, for networks.

FISU   Fill-In Signaling Unit, 'idle' packet that carries ACKs as sequence numbers (SS7).

FITL   Fiber In The Loop, optical technology from CO to customer premises.

FIX   Federal Internet Exchange, point of interconnection for U.S. agency research networks.

FMBS   Frame Mode Bearer Service, FR on ISDN.

FO   Fiber Optic, based on optical cable.

FOTP   Fiber Optic Test Procedure.

FOTS   Fiber Optic Terminal System, mux or CO switch interface.

FPDU   FTAM PDU.

FPDU   Frame relay Protocol Data Unit (I.122).

FR   Frame Relay, interface to simplified packetized switching network (I.122, T1.617).

FRAD   Frame Relay Access Device (Assembler/ Disassembler), functions like a PAD or router
             for frame relay networks.

FRBS   FR Bearer Service, newer name for FMBS.

VOICE OVER FRAME RELAY   **141**

FRF   Frame Relay Forum.

FRF-TC   FRF Technical Committee, writes implementation agreements.

FRS   Frame Relay Switch or Service.

FRSE   FR Service Emulation, as by IWF over ATM.

FRMR   Frame Reject, LLC response to error that cannot be corrected by ARQ, may cause reset or disconnect (Layer 2).

FS   Failed Second, now called UAS.

FSK   Frequency Shift Keying, modem method where carrier shifts between two fixed frequencies in voice range.

FSN   Forward Sequence Number, sent sequence number of this SU/packet (SS7).

FT-1   Fractional T-1, digital capacity of N x 64 kbit/s but usually less than $^1/_2$ a T-1.

FTAM   File Transfer, Access, and Management; an OSI layer-7 protocol for LAN interworking (802).

FTP   File Transfer Protocol (TCP/IP).

FTTC   Fiber To The Curb (Cabinet), local loop is fiber from CO to just outside CP, wire into CP.

FUNI   Frame-based UNI, serial interface to ATM network.

FX   Foreign Exchange, not the nearest CO. FX line goes from a CO or PBX to beyond its normal service area.

FXO   Foreign Exchange, Office; an interface at the end of a private line connected to a switch.

FXS   Foreign Exchange, Subscriber (or Station); an interface at the end of an FX line connected to a telephone, etc.

G

G   Ground, lead at VG port.

G3   Group 3, analog facsimile standard at up to 9.6 kbit/s.

G4   Group 4, digital facsimile standard at 56/64 kbit/s.

GA   Group Address, for multicasting.

Gbit/s   Giga bits per second, billions (10^9) per second.

GCRA   Generic Cell Rate Algorithm, how an ATM entity measures/controls negotiated service usage.

GFC   Generic Flow Control, first half-byte in ATM header at UNI.

GOSIP   Government OSI Profile, suite of protocols mandated for US Federal and U.K. contractors; -T = Transport model; -A = Application model.

.gov   Government, Internet domain name.

GPS   Global Positioning System, satellites that report exact time.

GS   Ground start, analog phone interface.

GSM   Group Special Mobile (Global System for Mobile communications), CEPT standard on digital cellular.

GSTM   General Switched Telephone Network, CCITT term to replace PSTN after 1990's privatizations.

GUI   Graphical User Interface.

H

H	Halt, line state symbol (FDDI).
H	Henry, unit of inductance.
Hx	High-speed bearer channels (ISDN):
H0	384 kbit/s.
H1	Payload in a DS-1 channel.
H11	1.536 Mbit/s (N. Amer.).
H12	1.920 Mbit/s (CEPT areas).
H2	Payload of a DS-3 channel.
H21	32.768 Mbit/s (CEPT).
H22	43.008 Mbit/s, the payload of 28 T-1s (N. A.), to 41.160 Mbit/s, if including 18 more DS-0s from the DS-3 overhead (N. Amer.)
H3	Would have been 60-70 Mbit/s, but left undefined for lack of interest.
H4	135.168 Mbit/s (88 T-1s)
HCDS	High Capacity Digital Service, Bellcore T-1 specification.
HCI	Host Command Interface, Mitel's PHI.
HCM	High Capacity Multiplexing, 6 channels of 9600 in a DS-0.
HCS	Header Check Sequence, CRC on header fields only, not on info; HEC (ATM).
HCV	High Capacity Voice, 8 or 16 kbit/s scheme.
HDB3	High Density Bipolar 3-zeros, line coding for 2 Mbit/s lines replaces 4 zeros with BPV (CEPT).
HDLC	High-level Data Link Control, layer-2 full-duplex protocol.
HDSL	High bit-rate Digital Subscriber Line, Bellcore Standard for way to carry DS-1 over local loops without repeaters.
HDT	Host Digital Terminal, CO end of multiplexed local loop (see RDT).
HDTV	High Definition Television, double resolution TV image and candidate application for broadband networks.
HE	Header Extension, a 12-octet field for various information elements (SMDS).
HEC	Header Error Control, ECC in ATM cell for header, not data. (See HCS)
HEL	Header Extension Length, the number of 32-bit words in HE (802.6).
HEPNET	High Energy Physics Network, international R&D net.
HFC	Hybrid Fiber-Coax, fiber from CO to neighbor cabinet, then coax to CP.
HIPPI	HIgh-speed Peripheral Parallel Interface, computer channel simplex interface clocked at 25 MHz; 800 Mbit/s when 32 bits wide, 1.6 Gbit/s when 64 bits.
HLM	Heterogeneous LAN Management, OSI NMS protocol specification without layers 3-6, developed by IBM and 3Com to save memory in workstations.
HLPI	High Layer Protocol Identifier; field in L3-PDU, included in SMDS to align with DQDB (802.6).
HOB	Head Of Bus, station and function that generates cells or slots on a bus (DQDB).
HOPS	Horizontally Oriented Protocol Structure, proposal for high performance interfaces at broadband rates.

HPR	High Performance Routing, a form of dynamic call routing in the PSTN.
HQ	Headquarters, network central site.
HRC	Hybrid Ring Control, TDM sublayer at bottom of data link (2) that splits FDDI into packet- and circuit switched parts.
HSSI	High Speed Serial Interface, of 600 or 1200 Mbit/s.
HSPS	High Speed Peripheral Shelf.
HTML	HyperText Mark-up Language, text commands to create WWWeb pages.
HTTP	HyperText Transfer Protocol.
Hz	Hertz, frequency, cycles/second.

I

I	I class central office switch is not in HPR network.
I	Idle, line state symbol (FDDI).
I	Information, type of layer 2 frame that carries user data.
IA	Implementation Agreement, based on a subset of CCITT or ISO standards without options to ensure interoperability.
Ia	Interface point a, the network-side port of TE-1, NT-1, or NT-2 (ISDN PRI)
Ib	Interface point b, the user-side port of TE-1, NT-1, or NT-2 (ISDN PRI).
IA5	International Alphabet #5, coding for signaling information (ISDN).
IAB	Internet Activities Board, defines LAN standards like SNMP.
IACS	Integrated Access and Cross-connect System, AT&T box with DACS and mux functions via packet switching fabric.
IAD	Integrated Access Device, CPE that supports multiple services on public/private net.
IAM	Initial Address Message, call request packet (SS7).
IBR	Intermediate Bit Rate, between 64 and 1536 kbit/s; fractional T-1 rates.
IC	Integrated Circuit.
ICA	International Communications Association, a users group.
ICCF	Industry Carriers Compatibility Forum.
ICIP	InterCarrier Interface Protocol, connection between two public networks.
ICMP	Internet Control Message Protocol, reports to a host errors detected in a router by IP.
ICR	Initial Cell Rate.
IDF	Intermediate Distribution Frame.
IDLC	Integrated Digital Loop Carrier, combination of RDT (remote mux), transmission facility, and IDT to feed voice and data into a CO switch.
IDT	Integrated Digital Terminal, M24 function in a CO switch to terminate a T-1 line from RDT.
IE	Information Element, part of a message; e.g. status of one PVC in a report.
IEC	Inter-Exchange Carrier, a long distance company, carries traffic between LATA's.
IEC	International Electrotechnical Commission, standards body.
IEEE	Institute of Electrical and Electronics Engineers, Inc.; engineering society; one of the groups which set standards for communications.

IETF	Internet Engineering Task Force, adopts RFCs.
I/F	Interface.
IG	ISDN Gateway (AT&T).
IGOSS	Industry/Government Open Systems Specification, broader GOSIP.
IGP	Interior Gateway Protocol, IGRP.
IGRP	Interior Gateway Routing Protocol, learns best routes through large LAN internet (TCP/IP).
ILMI	Interim Local Management Interface, PVC management in ATM at UNI.
ILS	Idle Line State, presence of idle codes on optical fiber line (FDDI).
IMD	InterModulation Distortion.
IMPDU	Initial MAC PDU, the SDU received from LLC with additional header/trailer to aid in segmentation and reassembly (802.6).
IN	Intelligent Network.
INA	Integrated Network Access, multiple services over one local loop.
IND	Indication (OSI).
INE	Intelligent Network Element,
I/O	Input/Output.
IOC	Inter-Office Channel, portion of T-1 or other line between COs of the IXC.
IOC	Isdn Ordering Code, 1 or 2-letter code for a complete set of configuration parameters for ISDN BRI service.
IMPDU	Initial MAC Protocol Data Unit.
IP	Internet Protocol, connectionless datagram network layer (3) basis for TCP, UDP, and the Internet.
IPX	Internetwork Packet eXchange, Novell's networking protocol, based on XNS.
IR	InfraRed, light with wave length longer than red, like 1300 nm used over fiber.
IS	International Standard.
ISA	Industry Standard Architecture, the personal computer design based on IBM's AT model.
ISDN	Integrated Services Digital Network.
ISDN-UP	ISDN User Part, protocol from layer 3 and up for signaling services for users, Q.761-Q.766 (SS7).
ISDU	Isochronous Service Data Unit, upper layer packet from TDM or circuit-switched service (802.6).
ISI	Inter-Symbol Interference, source of errors where pulses (symbols) spread and overlap due to dispersion.
ISO	International Standards Organization, ANSI is US member.
ISR	Intermediate Session Routing, performs address swapping and flow control (APPN).
ISSI	Inter-Switching System Interface, between nodes in a public network, not available to CPE (e.g. SMDS to B-ISDN).
ISSIP	ISSI Protocol.
ISUP	ISDN User Part (SS7).
ITB	End of Intermediate Transmission Block, control byte in BSC.

VOICE OVER FRAME RELAY **145**

ITG	Integrated Telemarketing Gateway.
ITU	International Telecommunications Union, UN agency, parent of (former) CCITT, CCIR, etc.
ITUA	Independent T1 Users Association (dissolved, 1995).
ITU-RS	ITU-Radiocommunications Sector.
ITU-T	Short for ITU-TSS
ITU-TSS	ITU Telecommunications Standardization Sector, successor to CCITT.
IVR	Interactive Voice Response.
IWF	InterWorking Function, the conversation process between FR and X.25, FR and ATM, etc.
IWU	InterWorking Unit, protocol converter between packet formats like FR and ATM.
IXC	IntereXchange Carrier, a long distance phone company or IEC, as opposed to LEC.

J

J	Non-data character for starting delimiter (11000) in 4B/5B coding (802.6).
JB7	Jam Bit 7, force bits in position 7 within a DS-0 to 1 for 1's density.
JPEG	Joint Photographic Experts Group, part of ISO that defined digital storage format for still photos.
JTC1	Joint Technical Committee 1, of IEC and ISO.

K

K	Non-data character for starting delimiter (10001) in 4B/5B coding (802.6).
k	Kilo, prefix for 1000; 1000 bit/s; K = 1024 bytes when applied to RAM size.
K2	LOH byte (SONET).
kbit/s	Thousands of bits per second
KDD	Kokusai Denshin Denwa, Japan's international long distance carrier.
KG	Key Generator (Krypto Gear), encryption equipment from NSA.
kHz	Kilohertz, thousands of cycles per second

L

L2-PDU	Layer 2 Protocol Data Unit, fixed length cell (SMDS).
L3-PDU	Layer 3 Protocol Data Unit, a variable length packet at OSI level 3.
LADT	Local Area Data Transport, telco circuit on copper pair.
LAN	Local Area Network.
LANE	LAN Emulation, ATM specification for logical linking of stations.
LAP	Link Access Procedure (Protocol), layer 2 protocol for error correcting between master station and 1 or more slaves.
LAPB	LAP Balanced, HDLC layer 2 protocol for data sent between 2 peer stations; used into X.25 network, etc.

LAPD   Variant of LAPB for ISDN D channels.

LAPD+   LAPD protocol for other than D channels, e.g. B channels.

LAPM   LAP Modem, part of V.42 modem standard.

LAT   Local Area Transport, DECnet protocol for terminals.

LATA   Local Access and Transport Area, a geographic region. The LEC can carry all traffic within a LATA, but nothing between LATA's.

LBO   Line Build Out, insertion of loss in a short transmission line to make it act like a longer line.

LBRV   Low Bit Rate Voice, digital voice compressed encoded below 64 kbit/s.

LC   Local Channel, the local loop.

LCD   Liquid Crystal Display.

LCD   Loss of Cell Delineation, receiver can't find cells (ATM).

LCI   Logical Connection Identifier, short address in connection-oriented frame.

LCN   Logical Channel Number, form of PVC address in an X.25 packet.

LCP   Link Control Protocol, part of PPP.

LD-CELP   Low Delay CELP, voice compression with small processing delay (G.728).

LDC   LATA Distribution Channel, line between local CO and POP.

LDM   Limited Distance Modem.

LE   LAN Emulation (ATM).

LEC   LAN Emulation Client, end station function maps MAC to ATM address.

LEC   Local Exchange Carrier, a telco.

LECS   LAN Emulation Configuration Server, part of LANE (ATM).

LED   Light Emitting Diode, semiconductor used as light source in FO transmitters.

LEN   Low Entry Networking, most basic subset of APPN.

LEN   Local Exchange Node, CO switch of LEC.

LEO   Low Earth Orbiting satellite, option for global 'cellular' phones.

LGE   Loop- or Ground-start, Exchange; FXO analog voice interface.

LGS   Loop- or Ground-start, Subscriber; FXS analog voice interface.

LI   Length Indicator, field in VoFR subframe header.

LI   Link Identifier, address consisting of VPI and VCI (ATM)

LIDB   Line Identification Data Base (SS7).

LIV   Link Integrity Verification (FR).

LLB   Local Loop Back.

LLC   Logical Link Control, the upper sublayer of the OSI data link layer (layer 2).

LLC1   Connection oriented LLC.

LLC2   Connectionless LLC.

LM   Layer Management, control function for protocol.

LME   Layer Management Entity, the process that controls configuration, etc. (802.6).

LMI   Layer Management Interface, software at each OSI layer in SMDS, 802.6.

LMI   Local Management Interface, transport specification for frame relay that sets way to report status of DLCIs.

VOICE OVER FRAME RELAY   **147**

LOF   Loss Of Frame, condition where mux cannot find framing, OOF, for 2.5 sec.

LOFC   · Loss of Frame Count, number of LOFs.

LOP   Loss of Pointer, SONET error condition, like LOF.

LOS   Loss Of Signal, incoming signal not present (no received data).

LPC   Linear Predictive Coding, voice encoding technique.

LPDA   Link Problem Determination Aid, part of Netview NMS (SNA).

LS   Loop Start, analog phone interface.

LSAP   Link layer Service Access Point, logical address of boundary between layer 3 and LLC sublayer in 2 (802).

lsb   Least Significant Bit, position in data field with smallest value.

LSSU   Link Status Signaling Unit, control packet at layer 3 (SS7).

LSU   Line State Unknown, possible report from FDDI line state monitor.

LT   Loop (Line) Termination, in the CO on a BRI.

LTE   Line Terminating Equipment, SONET nodes that switch, etc. and so create or take apart an SPE (SONET).

LU   Logical Unit, upper level protocol in SNA.

LUNI   LAN UNI, specific type of LAN emulation (ATM).

LU6.2   Set of services that support program to program communications.

M

M   Maintenance, overhead bits in frames and superframes at BRI.

M   Million when used as prefix to abbreviation: Mbit/s.

M   'Mouth' lead on VG switch, sends signaling.

m   Milli (1/1000) when used as prefix: mm = millimeter

m   Meter (39.37 inches).

M13   Multiplexer between DS-1 and DS-3 levels.

M24   Multiplexer function between 24 DS-0 channels and a T-1, a channel bank.

M28   Same as M13, but different format, not compatible.

M44   Multiplexer function to put 44 ADPCM channels into one T-1; four bundles, each of one common signaling channel with 11 voice channels; transcoder or BCM.

M48   Multiplexer function to put 48 ADPCM channels into one T-1; signaling in each voice channel.

M55   ADPCM multiplexer that puts 55 voice channels in five bundles on an E-1.

mA   Milliampere, unit of electrical current, 1/1000 of an ampere.

MAAL   AAL Management.

MAC   Medium (Media) Access Control, the lower sublayer of the OSI data link layer.

MAN   Metropolitan Area Network, typically 100 Mbit/s.

MAP   Manufacturing Automation Protocol, for LAN's; closely related to TOP, and written MAP/TOP (802.4).

MAU	Media Access Unit, device attached physically to Ethernet cable (802.3).
MAU	Multiple (Multistation, Media) Access Unit, hub device in a TR LAN (802.5).
Mbit/s	Megabit (1,000,000 bits) per second. .
MBS	Maximum Burst Size, number of cells that may be sent at PCR without exceeding SCR (ATM).
MCC	Master Control Center, part of DEC's umbrella network management system, EMA.
MCF	MAC Convergence Function, how an SDU is framed into a packet (PDU), segmented, and loaded into cells (802.6).
MCP	MAC Convergence Protocol, segmentation and reassembly procedure to put MSDUs into cells (802.6).
MCPC	Multiple Channel Per Carrier, satellite connection method for point-point links.
MCR	Minimum Cell Rate.
MDDB	Multi-Drop Data Bridging, digital bridging of PCM encoded modem signals, equivalent to analog bridging.
MDF	Main Distribution Frame, large CO wire rack for low speed data and voice cross connects.
MDI	Medium Dependent Interface, link between MAU and cable (802 Layer 1).
MELP	Mixed Excitation Linear Prediction, voice encoding method for LBRV; DoD standard.
MF	Multi-Frequency, tone signaling on analog circuits between CO switches.
MFA	MultiFrame Alignment, code in time slot 16 of E-1 to mark start of superframe.
MFJ	Modified Final Judgment, court decision that split AT&T in 1984.
MHS	Message Handling System, OSI store and forward protocol.
MHz	Megahertz, million cycles per second.
MIB	Management Information Base, OSI defined description of a network for management purposes (SNMP, IP).
MIC	Media Interface Connector, dual-fiber equipment socket and cable plug (FDDI).
MID	Message IDentification, a sequence number shared by all L2-PDUs holding segments of one L3-PDU (SMDS, 802.6) or all segments of same frame (AAL3/4).
MIDI	Musical Instrument Digital Interface.
MIPS	Millions of Instructions Per Second, speed rating for computer.
MIS	Management Information Systems, dept. that runs the big computers.
MJU	Multipoint Junction Unit, a digital data bridge for DDS (DS-0B or 56 kbit/s), often part of a DACS.
MLHG	MultiLine Hunt Group, operation mode for ISDN terminal.
MLPPP	MultiLink PPP, protocol to split data stream over multiple channels.
MMFS	Manufacturing Messaging Format Standard, application protocol (MAP).
MML	Man-Machine Language, commands and responses understandable by both human and device being controlled.
MNP	Microcom Networking Protocol, error correcting protocol and compression in modems.
modem	MOdulate/DEModulate, modulate analog signal from digital data and reverse.
MOS	Mean Opinion Score, scale for voice quality from 5 (toll quality) to 1 (unusable) assigned by expert listeners.

MPA	Manufacturing Program Analysis, evaluation standard for vendor plants (TR411).
MPEG	Motion Picture Experts Group, part of ISO that defined digital video compression and file format.
MPDU	MAC PDU (802.6).
MPL	Maximum Packet Lifetime, number of hops allowed before packet is discarded.
MPMC	Multi-Peer Multicast, "N-way" mutual broadcasting of information.
MPOA	MultiProtocol Over ATM.
MRD	Manual Ring Down, VG leased line where caller presses button to ring.
MS	Management Service (SNA).
ms	Millisecond, 1/1000 second.
M/S	Master/Slave, relationship in a protocol where master always issues commands and slave only responds.
MSAP	MAC Service Access Point, logical address (up to 60 bits) of boundary between MAC and LLC sublayers (802).
msb	Most Significant Bit, high order bit in a data field.
MSDU	MAC Service Data Unit, data packet in LAN format; may be long and variable length before segmentation into cells.
MSS	MAN Switching System.
MSS	Maximum Segment Size, limit on TCP frames sent.
MSU	Message Signaling Unit, layer 3 packet (SS7).
MTA	Metallic Test Access, service point on equipment in CO.
MTBF	Mean Time Between Failures, average for one device.
MTBSO	Mean Time Between Service Outages.
MTP	Message Transfer Part, set of connectionless protocols at lower layer 3 and below (SS7); cf ISUP.
MTS	Message Telephone (Toll) Service, normal dial up phone service.
MTSO	Mobile Telephone Switching Office, CO that joins cellular and landline services.
MTTR	Mean Time To Repair.
MTU	Maximum Transmission Unit, largest PDU in IP.
MUX	Multiplexer.
mW	Milliwatt, unit of electrical power, 1/1000 watt.

N

N	N class central office has tandem switch that participates in HPR.
N	Digit from 2 to 9 inclusive.
n	Nano, prefix meaning 10-9 of the unit as nm = 10-9 meter.
N.A.	North America.
NAK	Negative Acknowledgment, protocol control byte indicating error.
NANP	North American Numbering Plan.
NAS	Network Applications Solutions, set of DEC APIs for communication.

NAU   Network Addressable Unit, addressable device or process running an SNA protocol.

NBS   National Bureau of Standards, now NIST.

N.C.   Normally Closed, switch contacts on at 'idle.'

NCB   Network Control Block, command packet in SNMP.

NCB   Network Control Block, transport protocol in LAN Manager (level 4).

NCC   Network Control Center.

NCI   Network Control Interface.

NCP   Network Control Point, for SDN and AT&T switched network.

NCP   Network Control Program, software for FEP in SNA; has FR interface after Ver. 7.1.

NCP   Network Control Protocol, part of PPP.

NCTE   Network Channel Terminating Equipment; first device at CP end of local loop; e.g., CSU.

NDF   New Data Flag, inversion of some pointer bits to indicate change in SPE position in STS frame (SONET).

NDIS   Network Driver Interface Specification.

NE   Network Element, device or all similar devices in a network.

NEBS   Network Equipment-Building System, Bellcore generic spec for CO equipment (TR63).

NECA   National Exchange Carriers Association.

NET3   EC standard for BRI.

NET5   EC standard for PRI.

NET33   EC standard for ISDN telephones.

NEXT   Near End Cross Talk, interference on 2-wire interfaces from sent signals leaking back into the receiver.

NFS   Network File System, protocol for file transfers on a LAN.

NFS   Network File Server, computer with shared storage, on a LAN.

NI   Network Interface; demarcation point between PSTN and CPE/CI.

NIC   Network Interface Card, add-in card for PC, etc. to connect to LAN.

NID   Network IDentification, field in network level header (MAP).

NISDN   Narrowband ISDN, access at T-1 or less.

NIST   National Institute of Standards and Technology, name change for National Bureau of Standards.

NIUF   North-american Isdn Users Forum, a group associated with NIST.

NLPID   Network Level Protocol ID, control field in frame header identifying encapsulated protocol.

NM   Network Management.

NME   NM Element.

NMOS   N-channel Metal Oxide Semiconductor, common IC type uses more power than CMOS.

NMP   Network Management Protocol.

NMS   Network Management System.

NMVT   Network Management Vector Transport (SNA).

NNI   Network-Network Interface, between two carriers or between carrier and private net-work (FR, ATM).

NNI	Network-Node Interface, point to point interface between two switches for SDH, SONET, or B-ISDN network.
N.O.	Normally Open, switch contacts off at 'idle.'
NOS	Network Operating System.
NPA	Numbering Plan Area, area code in phone number: NPA-NXX-5555.
NPDA	Network Problem Determination Application, fault isolation software for IBM hosts, part of NetView.
NPDU	Network PDU, layer 3 packet (OSI).
NPI	Numbering Plan Indicator, field in message with DN to specify local, national, or international call (ISDN).
NPSI	Network Packet Switching Interface, IBM software for packet connection to FEP (SNA).
NR	Number Received, control field sequence, tells sender the NS that receiver expects in next frame (Layer 2).
NREN	National Research and Education Network, U.S.
NRZ	Non-Return to Zero, signal transitions from positive to negative without assuming 0 value. See also DMC, AMI.
NRZI	NRZ Invert on ones, coding changes polarity to indicate '1' and remains unchanged for '0.'
ns	Nanosecond, 10-9 second.
NS	Network Supervision (ATM).
NS	Number Sent, sequence number of frame in its control field; determined by sender.
NSA	National Security Agency.
NSA	Non-Service Affecting, fault that does not interrupt transmission.
NSAP	Network Service Access Point, logical address of a 'user' within a protocol stack (ISDN).
NSC	Network Service Center, for SDN.
NSDU	Network Service Data Unit, basic packet passed by SCCP (SS7); also OSI.
NSFNET	National Science Foundation Network.
NSP	Network Services Part, reliable transport for signaling, MTP + SCCP.
NT-1	Network Termination 1, the first device on the CP end of the ISDN local loop (like the CSU).
NT-2	Network Termination 2, the second CP device, like the DSU (ISDN).
NTM	NT Test Mode, BRI control bit.
NTN	Network Terminal Number, address of terminal on data network, part of global address with DNIC (X.121).
NTSC	National Television Standards Committee, group and format they defined for U.S. TV broadcasting.
NTT	Nippon Telephone and Telegraph, the domestic phone company in Japan.
NUI	Network/User Interface.
NV	NetView, IBM's umbrella network management system.
NWT	Network Technology, Bellcore group.
NXX	Generic indication of exchange in phone number: NPA-NXX-5555.
NYSERnet	New York State Education and Research Network, part of NSFnet.

O

OAI   Open Applications Interface, Intecom's PHI.

OAM   Operations, Administration, and Maintenance.

OAM&P   Operations, Administration, Maintenance, & Provisioning, telco housekeeping.

OC-1   Optical Carrier level 1, SONET rate of 51.84 Mbit/s, matches STS-1.

OC-3   Optical Carrier level 3, SONET rate of 155.52 Mbit/s, matches STS-3.

OC-N   Higher SONET levels, N times 51.84 Mbit/s.

OCD   Out of Cell Delineation, receiver is searching for cell alignment (ATM).

OCR   Office Channel Repeater, OCU.

OCU   Office Channel Unit, "CSU" in the CO; also called OCR.

OCU-DP   OCU-Data Port, channel bank plug I/O to 4-wire local loop and CSU on CP to provide DDS.

ODI   Open Data-link Interface, driver interface, API for LAN cards.

OEM   Original Equipment Manufacturer.

OF   Optical Fiber.

OLTP   On Line Transaction Processing.

OMAP   Operations Maintenance and Administration Part, upper layer 7 protocol in SS7.

ONA   Open Network Architecture, FCC plan for equal access to public networks.

ONI   Operator Number Identification.

OOF   Out Of Frame, mux is searching for framing bit pattern.

OOS   Out Of Synchronization; multiplexers can't transmit data when OOS.

OPC   Origination signal transfer Point Code, address in SU of source of packet (SS7).

OPR   Optical Power Received, by a FO termination.

OPX   Off-Premises Extension, line from PBX to another site.

OR   Or, as in either/or, a logical device that outputs a 1 if any input is 1; a 0 only if all inputs are 0.

.org   Organization, first Internet domain for non-profits, schools, etc.

ORL   Optical Return Loss.

OS   Operating System, main software to run a CPU.

OS   Operations System, used by telco to provision, monitor, and maintain facilities.

OSF   Open Software Foundation.

OSI   Open Systems Interconnection, a 7-layer model for protocols defined by the ISO.

OSI/NMF   OSI Network Management Forum, standards group for NM protocols.

OSIone   Global organization to promote OSI standards.

OSI TP   OSI Transaction Processing, a protocol.

OSPF   Open Shortest Path First, standard routing protocol.

OSS   Operations (or Operational) Support System, used by telco to provision, monitor, and maintain facilities.

OTC   Operating Telephone Company, LEC.

OTC   Overseas Telephone Company, international carrier in Australia.

OTDR   Optical Time Domain Reflectometry (Reflectometer), method (tester) to locate breaks in OF.

OUI   Organizationally Unique Identifier, code for administrator of PIDs.

OW   Order Wire, DS-0 in overhead intended for voice path to support maintenance .

P

PA   Preamble, a period of usually steady signal ahead of a LAN frame, to set timing, reserve the cable, etc.

PA   Pre-Arbitrated, portion of traffic on DQDB MAN with assigned bandwidth, usually isochronous connections (802.6).

PABX   Private Automated Branch eXchange, electronic PBX.

PAD   Packet Assembler/Disassembler, device to convert between packets (X.25, etc.) and sync or async data.

PAL   Programmable Array Logic, large semi-custom chip.

PAM   Pulse Amplitude Modulation; used within older channel banks and at 2B1Q ISDN U interface.

PANS   Peculiar and Novel Services, phone services that go beyond POTS: switched data, ISDN, etc.

PAP   Password Authentication Protocol, encrypts passwords for security of dial-in access.

PARIS   Packetized Automated Routing Integrated System, fast switch developed by IBM.

PBX   Private Branch eXchange, small phone switch inside a company, manual or automatic.

PC   Path Control, level 3 in SNA for network routing.

PC   Personal Computer, often used as a data terminal.

PCB   Printed Circuit Board.

PCC   Page Counter Control (SMDS).

PCI   Peripheral Component Interconnect, Intel's advanced bus for personal computers.

PCI   Protocol Control Information (ATM).

PCM   Page Counter Modulus, SMDS header field.

PCM   Pulse Code Modulation, the standard digital voice format at 64 kbit/s.

PCN   Personal Communications Network, second generation cellular system.

PCMCIA   Personal Computer Memory Card International Association.

PCR   Peak Cell Rate, traffic parameter applied per VC, VP, or channel (ATM).

PCR   Preventive Cyclic Retransmission, error correction procedure that repeats packets whenever link bandwidth is available (SS7).

PCS   Personal Communications Service, low-power portable phones based on dense public network of small cells.

PDG   Packet Data Group, 12 octets in FDDI frame (outside of WBCs) not assignable to circuit-switched connections.

PDH   Plesiochronous Digital Hierarchy, present multiplexing scheme from T-1 to T-3 and higher; contrast with SDH.

PDN   Public Data Network; usually packetized.

PDS   Premises Distribution System, the voice and data wiring inside a customer office.

PDU   Protocol Data Unit, information packet (ADDR, CTRL, INFO) passed at one level between different protocol stacks (OSI).

pel   Picture Element, the smallest portion of a graphic image encoded digitally.

P/F   Poll/Final, bit in control field of LLC frames to indicate receiver must acknowledge (P) or this is last frame (F) (Layer 2).

PFT   Power Failure Transfer, protection switch.

PHF   Packet Handling Facility, packet switch for X.25 or FR service (ISDN).

PHI   PBX-Host Interface, generic term for link between voice switch and computer, c.f., SCAI.

PHY   PHYsical, layer 1 of the OSI model.

PI   Primary In, FO port that receives light from main fiber ring (FDDI).

PIC   Polyethylene Insulated Cable, modern phone wire.

PID   Protocol ID, codes (some allotted by CCITT) to identify specific protocols.

PIN   Positive-Intrinsic-Negative, type of semiconductor photo detector.

PIU   Path Information Unit, BIU plus the transmission layer frame header (SNA).

PL   Pad Length, number (0-3) of octets of 0s added to make Info field a multiple of 4 octets (802.6).

PL   Payload Length, field in VoFR subframe header.

PL   Physical Layer, level 1 in OSI model.

PL   Private Line, a dedicated leased line, not switched.

PLAR   Private Line Automatic Ring-down; see ARD.

PLB   Performance Loop Back, LB done at point of ESF performance function in CPE.

PLCP   Physical Layer Convergence Protocol (Procedure), part of PHY that adapts transmission medium to handle a given protocol sublayer (DQDB).

PLL   Phase Locked Loop, electronic circuit that recovers clock timing from data.

PLP   Packet Layer Protocol, at layer 3 like X.25.

PLS   Physical Link Signaling, part of Layer 1 that encodes and decodes transmissions, e.g. Manchester coding (IEEE 802).

PM   Performance Monitoring, function in ATM.

PMA   Physical Medium Attachment, electrical driver for specific LAN cable in MAU, separated from PLS by AUI (802.3).

PMA   Primary Market Area, metro area as served by MAN.

PMD   Packet Mode Data, ISDN call type.

PMD   Physical layer, Medium Dependent; a sublayer in layer 1 (below PLS) of LAN protocols; also PMA (802).

PMP   Point to MultiPoint, broadcast connection (ATM UNI).

PNNI   Private Network-Network Interface, between ATM switches in public and private networks.

PO   Primary Out, FO port that sends light into the main fiber ring (FDDI).

POF   Plastic Optical Fiber, for short distances rather than glass for long haul.

POH   Path OverHead, bytes in SDH for channels carried between switches over multiple lines and through DCCs .

POP   Point Of Presence; end of IXC portion of long-distance line at central office (Tel).

POS   Point of Sale.

POTS	Plain Old Telephone Service, residential type analog service.
PPDU	Presentation (layer) PDU (OSI).
ppm	Parts Per Million, 1 ppm = 0.0001%.
PPP	Point to Point Protocol, non-proprietary multi-protocol serial interface for WAN links.
pps	Packets Per Second, switch capacity.
pps	Pulses Per Second, speed of rotary dialing dial pulses.
PR	Page Reservation, SMDS header field.
PRC	Primary Reference Clock, GPS-controlled rubidium oscillator used as stratum 1 source.
PRA	Primary Rate Access, via PRI for ISDN.
PRBS	Pseudo-Random Bit Sequence, fixed bit pattern, for testing, that looks random but repeats.
PRI	Primary Rate Interface; 23B+D (T-1) or 30B+D (CEPT).
PRM	Protocol Reference Model.
PROM	Programmable Read Only Memory; non-volatile type chip.
PRS	Primary Rate Source, stratum 1 clock.
PS	Power Status, 2-bit control field at BRI.
PS	Presentation Services, level 6 of SNA.
PSC	Public Service Commission, telecom regulator in many states, also PUC.
PSDN	Public Switched Data Network, national collection of interconnected PSDSs.
PSDS	Public Switched Digital Service, generic switched 56K intra-LATA service.
PSI	Primary Subnet Identifier, part of address in network level header (MAP).
PSK	Phase Shift Keying, modem modulation method.
PSN	Packet Switched Network.
PSN	Public Switched Network.
PSPDN	Packet Switched Public Data Network.
PSTN	Public Switched Telephone Network, the telco-owned dial-up network.
PT	Payload Type, field in frame or cell header.
PTAT	Private Trans-Atlantic Telephone, cable from US to U.K., Ireland, and Bermuda.
PTE	Path Terminating Equipment, SONET nodes on ends of logical connections.
PTI	Payload Type Identifier, control field in ATM header.
PTT	Postal, Telephone, and Telegraph authority; a monopoly in most countries.
PU	Physical Unit, SNA protocol stack that provides services to a node and to less intelligent devices attached to it.
PU2	Cluster controller or end system.
PU4	Front End Processor.
PUB	AT&T technical PUBlication, Bell System de facto standard, most from before divestiture.
PUC	Public Utilities Commission, state body that regulates telephones, also PSC.
PVC	Permanent Virtual Circuit (Connection), assigned connection over a packet, frame, or cell network, not switchable by user.
PVN	Private Virtual Network, VPN.
PWB	Printed Wiring Board, PCB.

Q

Q	Quiet, line state symbol (FDDI).
QA	Queued Arbitrated, portion of packet traffic that contends for bandwidth (DQDB).
QAM	Quadrature Amplitude Modulation, high speed modem, also used in CAP.
QFC	Quantum Flow Contrl, way to manage ABR service (ATM) based on buffer usage in switches.
QLLC	Qualified Logical Link Control, a frame format (SNA).
QoS	Quality of Service, performance specification for network..
QPSX	Queued Packet Synchronous eXchange, old name for DQDB; QPSX Systems Inc. originated it in Australia.
Q.921	CCITT recommendation for level 2 protocol in signaling system 7.
Q.931	CCITT recommendation for level 3 protocol in signaling system 7.

R

R	Interface reference point in the ISDN model to pre-ISDN phone or terminal.
R	Red alarm bit in synch byte (TS 24) of T1DM (Tel).
R	Reserved, bit or field in frame not yet standardized, not to be used.
R	Ring, one of the conductors in a standard twisted pair, 2-wire local loop (the one connected to the 'ring,' the second part of a phone plug) or the DTE-to-DCE side of a 4-wire interface.
R1	Ring, or R lead of the DCE-to-DTE pair in a 4-wire interface.
RACE	Research for Advanced Communications in Europe, program to develop broadband.
RACF	Remote Access Control Facility, security program (SNA).
RAI	Remote Alarm Indication, (yellow alarm) repeating pattern of 8 ones and 8 zeros in EOC of ESF T-1 line (also in ATM).
RAID	Redundant Array of Inexpensive Disks.
RAM	Random Access Memory; volatile chip.
RARE	Reseaux Associes pour la Recherche Europeene, European Organization of Research Networks.
RARP	Reverse ARP, Internet protocol to let diskless workstation learn its IP address from a server (see BOOTP).
RBHC	Regional Bell Holding Company, one of the seven "baby Bells."
RBOC	Regional Bell Operating Company, one of about 22 local telephone companies formerly part of Bell System.
RBS	Robbed Bit Signaling, in PCM.
RD	Receive Data, lead on electrical interface.
RD	Request Disconnect, secondary station unnumbered frame asking primary station for DISC (Layer 2).
RDA	Remote Database Access, service element (OSI).

RDT	Remote Digital Terminal, advanced channel bank functionality on fiber or copper loop.
REJ	Reject, S-format LLC frame acknowledges received data units while requesting retransmission from specific errored frame (Layer 2).
REL	RELease, signaling packet on disconnect (SS7).
RELC	Release Complete, packet to acknowledge disconnect (DSS1 and SS7).
RELP	Residually-Excited Linear Predictive Coding, voice encoding scheme (8-16 kbit/s).
REN	Ringer Equivalent Number, the load presented to CO line during ringing, compared to one analog phone.
REQ	Request (OSI).
RF	Radio Frequency.
RFC	Request For Comment, documents that are modified then adopted by IETF as Internet standards.
RFH	Remote Frame Handler, FR switch or network accessed over CS links.
RFI	Radio Frequency Interference.
RFP	Request For Proposal.
RFT	Remote Fiber Terminal, equivalent to SLC96.
RH	Request/response Header, 3 bytes added to user data in format for first upper layer frame (SNA).
RHC	Regional Holding Company, one of the 7 telco groups split from AT&T in 1984, see RBOC.
RI	Ring Indicator, digital lead on modem tells DTE when call comes in (phone rings).
RI	Routing Indicator, bit in LAN packet header to distinguish transparent- from source-routed packets.
RIM	Request Initialization Mode, layer 2 supervisory frame.
RIP	Routing Information Protocol, method for routers to learn LAN topology (TCP/IP).
RISC	Reduced Instruction Set Computer.
RJ	Registered Jack, connector for UNI; RJ11 is standard phone, RJ45 for DDS and terminal, RJ48 for T-1.
RJE	Remote Job Entry, one form of BSC.
RL	Ring Latency, time for empty token to traverse full ring with no load (FDDI).
RM	Rate Management, flow control (cell in ATM connection).
RM	Reference Model.
RMN	Remote Multiplexing Node.
rms	Root Mean Square, form of V average related to power in a.c. circuits.
RN	Redirecting Number, DN of party that forwarded a call via the network (ISDN).
RNR	Receiver Not Ready, S-format LLC frame acknowledges received data units but stops sender temporarily (Layer 2 HDLC).
RO	Receive Only.
ROH	Receiver Off Hook, signal from CO switch that finds line off-hook but not in use; "howler."
ROLC	Routing Over Large Clouds, IETF study.
ROM	Read Only Memory; non-volatile chip; applied to CD holding data.
ROSE	Remote Operation Service Element (OSI).

RPOA	Recognized Private Operating Agency, X.25 interexchange carrier (ISDN).
RR	Receive Ready, S-format LLC frame acknowledges received data units and shows ability to receive more (Layer 2).
RS	Radiocommunications Sector, part of ITU, 1993 successor to CCIR.
RSET	Reset, layer 2 supervisory frame to zero counters.
RSL	Request and Status Link, same as PHI or SCAI.
RSP	Response (OSI).
RSU	Remote Switch Unit, multiplexing equipment outside CO that serves a CSA.
RT	Remote Terminal, CP end of multiplexed access loop, a mux.
RTS	Request To Send; lead on terminal interface.
RTS	Residual Time Stamp, control information in ATM to support CBR service.
RTT	Round Trip Time, twice total transmission latency.
RTU	Remote Terminal (Test) Unit.
RU	Request/response Unit, unframed block of up to 256 bytes of user data (SNA).
RVI	Reverse Interrupt, positive ACK that lets station take control of a BSC line.
RZ	Return to Zero; signal pauses at zero voltage between each pulse, when making zero crossings.

S

S	Status, signaling bit in CMI.
S	ISDN interface point between TA and NT-2.
S	Supervisory frame, commands at LLC level: RR, RNR, REJ, SREJ (Layer 2).
s	Second (unit of time).
S0	European notation for BRI.
S2	European notation for PRI (30B+D).
SA	Source Address, field in frame header (802).
SA	Synchronous Allocation, time allocated to FDDI station for sending sync frames (802.6).
SAA	Systems Application Architecture, compatibility scheme for communications among IBM computers.
SAAL	Signaling ATM Adaption Layer, for Q.2931 messages.
SABM	Set Asynchronous Balanced Mode, connection request between HDLC controllers or LLC entities (Layer 2).
SABME	SABM Extended, uses optional 16-bit control fields.
SAFER	Split Access Flexible Egress Routing, service at one site from two toll offices over separate T-1 loops (AT&T).
SAI	S/T Activity Indicator, BRI control bit.
SAP	Service Access Point, logical address of a session within a physical station, part of a header address at an interface between sublayers (802).
SAP	Service Advertising Protocol, periodic broadcast by LAN device (Netware).
SAPI	Service Access Point Identifier, part of address between layers in protocol stack; e.g., subfield in first octet of LAP-D address.

SAR   Segmentation And Reassembly, protocol layer that divides packets into cells.

SAR-PDU   SAR Protocol Data Unit, segment of CS-PDU with additional header and possibly a trailer (e.g., a cell in ATM).

SARM   Set Asynchronous Response Mode, unnumbered frame connection request (layer 2 HDLC).

SARME   SARM Extended, uses optional 16-bit control field.

SARTS   Special Access Remote Test System, the way telcos test leased lines.

SAS   Single-Attached Station, FDDI node linked to network by 2 optical fibers (vs. DAS).

sB   Signal Battery, second lead to balance M lead in E&M circuit.

SBC   Sub-Band Coding, compressing voice into multiple bit streams.

SCADA   Supervisory Control and Data Acquisition, in nets for oil and gas producers.

SCAI   Switch-to-Computer Applications Interface, link between host CPU and voice switch to integrate applications; also PHI and RSL.

SCAMP   Single-Channel Anti-jam Man-Transportable, DoD project for LBRV terminal.

SCCP   Signaling Connection Control Part, upper layer 3 protocol (SS7).

SCIL   Switch Computer Interface Link, PHI by Aristacom.

SCP   Service Control Point, CPU and database linked to SS7 that supports carrier services (800, LIDB, CLASS).

SCPC   Single Channel Per Carrier, analog satellite technology (telephony).

SCR   Sustainable Cell Rate, traffic parameter (ATM).

SD   Starting Delimiter, unique symbol to mark start of LAN frame (JK in FDDI, HDLC flag, etc.).

S/D   Signal to Distortion ratio.

SDDN   Software Defined Data Network, virtual private network built on public data net.

SDH   Synchronous Digital Hierarchy, digital multiplexing plan where all levels are synched to same master clock,

SDLC   Synchronous Data Link Control; a half-duplex IBM protocol based on HDLC.

SDM   Subrate Digital (Data) Multiplexing, a DDS service to put multiple low-speed channels in a DS-0; also Multiplexer.

SDN   Software Defined Network.

SDS   Switched Digital Service, generic term for carrier function.

SDSL   Symmetric DSL, DSL with same bit rate in both send and receive directions.

SDU   Service Data Unit, information packet or segment passed down to become the payload of the adjacent lower layer in a protocol stack.

SEND   clear to Send, signaling bit in CMI.

SEP   Signaling End Point.

SES   Severely Errored Second, interval when BER exceeds 10^{-3}, >319 CRC errors in ESF, frame slip, or alarm is present.

SEV   Self-Excited Vocoder, form of CELP voice compression.

SF   Single Frequency; form of on/off-hook analog signaling within telcos.

SF   Subfield (SNA).

SF   Super Frame, 12 T-1 frames.

SFET   Synchronous Frequency Encoding Technique, a way to send precise isoc clocking rate as a delta from system clock.

sG Signal Ground, second lead to balance E lead in E&M signal circuit.

SHR Self-Healing Ring, topology can survive one failure in line or node (802.6, etc.).

SI Sequenced Information, LAP-D frame type.

SI Secondary In, FO port that receives light from secondary fiber ring (FDDI).

SIF Signaling Information Field, payload of a signaling packet or MSU (SS7).

SIM Set Initialization Mode, layer 2 supervisory frame.

SIO Service Information Octet, field in MSU used to identify individual users (SS7).

SIP SMDS Interface Protocol.

SIPO Signaling Indication Processor Outage, alarm on failure of processor that receives sig-naling packets ("indications") (SS7).

SIR Sustained Information Rate, average throughput; basis for SMDS access class.

SIT Special Information Tone, audible signal (often three rising notes) preceding an announcement by the network to a caller.

SITA Societe Internationale de Telecommunications Aeronautiques, operator of worldwide air-line network.

SIVR Speaker Independent Voice Recognition.

SLC Subscriber Loop Carrier, usually digital loop system.

SLIC Subscriber Line Interface Card (Circuit), on a switch.

SLIP Serial Line Internet Protocol, older PPP for IP only.

SLS Signaling Link Selection, field in routing label of SU that keeps related packets on same path to preserve delivery order (SS7).

SMAP Systems Management Application Process, all the functions at layer 7 and above to monitor and control the network (SS7).

SMB Server Message Block, a LAN client-server protocol.

SMDR Station Message Detail Recording, keeping list of all calls from each phone, usually by PBX or computer.

SMDS Switched Multi-megabit Data Service, offered on a MAN by a carrier; service mark of Bellcore.

SME Subject Matter Expert.

SME System Management Entity, process in ATM hardware that supports remote NMS.

SMF Single Mode Fiber, thin strand that supports only one transmission mode for low dis-persion of optical waves.

SMP Simple Management Protocol, newer and more robust than SNMP.

SMR Specialized Mobile Radio, for fleet management and dispatching.

SMT Station ManagemenT, NMS for FDDI.

SMTP Simple Mail Transfer Protocol (TCP/IP).

S/MUX Workstation software to allow UNIX daemons to talk to SNMP manager station.

SN Sequence Number, transmission order of frames or cells within channel or logical connection.

SNA SDH Network Aspects, evolving standards for VC payloads and network management (SDH).

SNA Systems Network Architecture, IBM's data communication scheme.

SNADS   SNA Distribution Services, communication architecture for electronic mail and other applications.

SNAP   Sub-Network Access Protocol, identifies encapsulated protocol and user (802.1, ATM).

SNI   Subscriber-Network Interface, the demark point.

SNMP   Simple Network Management Protocol, started in TCP/IP, but extending to many LAN devices (Layer 4-5).

SNP   Sequence Number Protection, CRC & parity calculated over SN field in header (AAL-1).

SNR   Signal to Noise Ratio, in dB.

SNRM   Set Normal Response Mode, unnumbered command frame (layer 2).

SNRME   SNRM Extended, uses optional 16-bit control field.

SO   Secondary Out, FO port that sends light into the secondary fiber ring (FDDI).

SO   Serving Office, central office where IXC has POP.

SOH   Section OverHead, bytes in SDH for channels carried through repeaters between line terminations like DCC or switch.

SOH   Start of Header, control byte in BSC.

SOHO   Small Office Home Office.

SONET   Synchronous Optical Network.

SP   Structure Pointer, field in AAL-1 cell (ATM).

SPAG   Standards Promotion and Applications Group, has same function as COS.

SPCS   Stored Program Controlled Switch, CO switch (analog or digital) controlled by a computer.

SPDU   Session (layer) PDU (OSI).

SPE   Synchronous Payload Envelope, data area in SONET/STS/SDH format, with POH.

SPF   Shortest Path First, LAN router protocol that minimizes some measure (delay) and not just "hops" between nodes.

SPID   Service Profile IDentifier, DN or DN plus unique extension (ISDN in N.A.).

SQPA   Software Quality Program Analysis, Bellcore process to evaluate vendors.

SREJ   Selective REJ, layer 2 frame that requests retransmission of one specific I frame.

SRL   Singing Return Loss.

SRT   Source Routing Transparent, variation of source routing combined with spanning tree algorithm for bridging (802).

SS7   Signaling System 7, CCS within PSTN; replaced CCIS or SS6.

SSA   Systems Applications Architecture, SNA plan to allow programs on different computers to communicate.

SSAP   Source Service Access Point, field in LLC frame header to identify the sending session within a physical station (802).

SSCF   Service Specific Coordination Function, maps SSCOP functions to lower layer (Q.2130, ATM).

SSCOP   Service Specific Connection Oriented Protocol, provides assured transport for Q.2931 PDUs (ATM).

SSCP   System Services Control Point, host software that controls SNA network.

SSCS   Service Specific Convergence Sublayer (ATM).

SSM   Single Segment Message, frame short enough to be carried in one cell.

SSN	SubSystem Number, local address of SCCP user (SS7).
SSP	Service Switching Point (ISDN).
ST	Stream, network layer protocol for very high speed connections.
STDM	Statistical Time Division Multiplexer.
STE	Secure Terminal Equipment.
STE	Section Terminating Equipment, SONET repeater.
STEP	Speech and Telephony Environment for Programmers, Wang's PHI.
STM	Synchronous Transfer Mode, one of several possible formats for SONET and BISDN.
STM-1	Synchronous Transport Module-1, smallest SDH bandwidth; = 155.52 Mbit/s, STM-n = n x 155.52 Mbit/s.
STP	Shielded Twisted Pair, telephone cable with additional shielding for high speed data and LANs.
STP	Signal Transfer Point, packet switch for SS7.
STS-1	Synchronous Transport Signal, level 1; electrical equivalent of OC-1, 51.84 Mbit/s.
STS-N	Signal in STS format at N x 51.84 Mbit/s.
STSX-n	Interface for cross-connect of STS-n signal that defines STS-n.
STX	Start of Text, control byte in BSC.
SU	Segmentation Unit, info field of L2-PDU, <= 44 octets of a L3-PDU (SMDS, 802.6).
SU	Signaling Unit, layer 3 packet (SS7).
SV	Subvector, part of NMVT (SNA).
SVC	Switched Virtual Circuit (Connection), temporary logical connection in a packet/frame network.
SVD	Simultaneous Voice and Data.
SWG	SubWorking Group, part of a technical committee or forum.
SWIFT	SWItched Fractional T-1, telco service defined by Bellcore, includes full T-1.
SWIFT	Society for Worldwide Interbank Financial Telecommunications, global funds transfer network of 2000 banks.
SW56	Switched 56 kilobit/s, digital dial up service.
SYN	Synchronization character, 16h ASCII.
sync	Synchronous.
SYNTRAN	Synchronous Transmission, byte aligned format for an electrical DS-3 interface.

T

T	Interface between NT-1 and NT-2 (ISDN).
T	Non-data character in 4B/5B coding, ending delimiter (802.6).
T	Measurement interval, seconds, = Bc/CIR.
T	Tip, one of the conductors in a standard twisted pair, 2-wire local loop (the wire connected to the 'tip' of a phone plug) or one of the DTE-to-DCE pair of a 4-wire interface.
T	Transparent, no robbed bit signaling in D4/ESF format.
T-1	Transmission at DS-1, 1.544 Mbit/s.

T1	The standards committee responsible for transmission issues in US, corresponds to ETSI (Europe) and the Telecommunications Technology Committee (Japan).
T1	Tip or T lead of the DCE-to-DTE pair in a 4-wire interface.
T1DM	T-1 Data Multiplexer, brings DS-0Bs together on a DS-1 (Tel).
T1D1	TSC of T1 for BRI U interface.
T1E1	TSC of T1 for SNI.
T1M1	TSC of T1 for NMS and OSS.
T1Q1	TSC of T1 for ADPCM, voice compression, etc.
T1S1	TSC of T1 for ISDN bearer services.
T1X1	TSC of T1 for SONET and SS7.
TA	Technical Advisory, a Bellcore standard in draft form, before becoming a TR.
TA	Terminal Adapter, matches ISDN formats (S/T) to existing interfaces (R) like V.35, RS-232.
TABS	Telemetry Asynchronous Block Serial, M/S packet protocol used to control network elements and get ESF stats.
TAC	Technical Assistance Center, network help desk.
TAPI	Telephony Applications Programming Interface.
TAPS	Test and Acceptance Procedures, telco document for equipment installation and set up.
TASI	Time Assigned Speech Interpolation; analog voice compression comparable to DSI and statistical multiplexing of data.
TAT	Trans-Atlantic Telephone, applied to cables, as TAT-8.
TAXI	100 Mbit/s interface to ATM switch.
TBD	To Be Determined, appears often in unfinished technical standards.
TC	Terminating Channel; local loop.
TC	Transport Connection.
TC	Transmission Control, level 4 in SNA.
TC	Trunk Conditioning, insertion of various signaling bits in A and B positions of DS-0 during carrier failure alarm condition.
TCA	TeleCommunications Association.
TCA	Threshold Crossing Alert, alarm that a monitored statistic has exceeded preset value.
TCAP	Transaction Capabilities Application Part, lower layer 7 of SS7.
TCC	Telephone Country Code, part of dialing plan.
TCP/IP	Transmission Control Protocol (connection oriented with error correction) on Internet Protocol (a connectionless datagram service).
TD	Transmit Data.
TDD	Telecom Device for the Deaf, Teletype machine or terminal with modem for dial-up access.
TDM	Time Division Multiplexing (or Multiplexer).
TDMA	Time Division Multiple Access, stations take turns sending in bursts, via satellite or LAN.
TDS	Terrestrial Digital Service, MCI's T-1 and DS-3 service.
TDSAI	Transit Delay Selection And Indication, way to negotiate delay across X.25 bearer service (ISDN).
TE	Terminal Equipment, any user device (phone, fax, computer) on ISDN service; TE1 supports native ISDN or B-ISDN formats (S/T interface); TE2 needs a TA.

TEI	Terminal Endpoint Identifier, subfield in second octet of LAP-D address field (ISDN).
TEST	Test command, LLC UI frame to create loopback (Layer 2).
TFT	Thin-Film Transistor, pixel in display panel.
TFTP	Trivial File Transfer Protocol, simpler than FTP.
TG	Transmission Group, one or more links between adjacent nodes (SNA).
TH	Transmission Header, 2 bytes in framing format for layer 4 protocol (SNA).
TIA	Telecommunications Industry Association, successor to EIA, sets some comms standards.
TIFF	Tagged Image File Format, for graphics files.
TIRKS	Trunk Inventory Record Keeping System, telco computer to track lines.
TIU	Terminal Interface Unit, CSU/DSU or NT1 for Switched 56K service that handles dialing. 61330
TLA	Three Letter Acronym.
TLI	Transport Level Interface, for UNIX.
TL1	Transaction Language 1, to control network elements (TR482); CCITT's form of MML.
TLP	Transmission Level Point, related to gain (or loss) in voice channel; measured power - TLP at that point = power at 0 TLP site.
TM	Traffic Management (ATM).
TMN	Telecommunications Management Network, a support network to run a SONET network.
TMS	Timing Monitoring System.
TN	TelNet, remote ASCII terminal emulation (TCP/IP); also TN-3270 for SNA over Ethernet.
TN	Transit Network, IEC (ISDN).
TN3270	Remote emulation of IBM 3270 terminal.
TO	Transmit Only; audio plug for a channel bank without signaling.
TOA	Type of Address, 1-bit field to indicate X.121 or not (X.25).
TON	Type Of Number, part of ISDN address indicating national, international, etc.
TOP	Technical and Office Protocol; for LAN's.
TOPS	Task Oriented Procedures, telco document for equipment operation and maintenance.
TOS	Type Of Service, connection attribute used to select route in LAN (SPF).
TP	Transaction Processing, work of a terminal on-line with a host computer.
TPEX	Twisted Pair Ethernet Transceiver.
TP-N	Transport Protocol of Class N (N=0 to 4), OSI layer 4.
TP-0	Connectionless TP (ISO 8602).
TP-4	Connection oriented TP (ISO 8073).
TPDU	Transport Protocol Data Unit (OSI).
TPF	Transaction Processing Facility, IBM host software for OLTP.
TPSE	Transport Processing Service Element (OSI).
TR	Technical Reference (Requirement), a final Bellcore standard.
TR	Token Ring, a form of LAN.
TS	Terminal Server, allows async terminal to talk over a LAN (Telnet/IP, e.g.)
TS	Time Slot, DS-0 channel in T-1, PRI, etc.
TS	Transaction Services, top level (7) of SNA protocol stack, on top of LU 6.2.

VOICE OVER FRAME RELAY **165**

TS   Transport Service (OSI).

TSAPI   Telelphony Server API.

TSB   Telecommunications Standardization Bureau, formed by ITU in 1993 from merger of CCITT and CCIR.

TSC   Technical Subcommittee, for standards setting.

TSDU   Transport Service Data Unit (OSI).

TSI   Time Slot Interchange(r); method (device) for temporarily storing data bytes so they can be sent in a different order than received; a way to switch voice or data among DS-0s (DACS).

TSS   Telecommunications Standardization Sector, a variant on TSB.

TSY   Technology Systems, Bellcore group renamed Network Technology (NWT).

TTC   Telecommunications Technology Committee, Japanese standards body.

TTL   Transistor-Transistor Logic; signals between chips.

TTR   Timed Token Rotation, type of token passing protocol (FDDI).

TTRT   Target Token-Rotation Time, expected or allowed period for token to circulate once around ring (802.4, 802.6).

TTY   Teletypewriter.

TU   Tranceiver Unit, active device at end of DSL.

TU   Tributary Unit, virtual container plus path overhead (SDH).

TUC   Total User Cells, count kept per VC while monitoring, field in OAM cell.

TUG   TU Group, one or more TUs multiplexed into a larger VC (SDH).

TUP   Telephone Users Part, ISDN signaling based on MTP without SCCP, used outside N.A. only.

TWX   TeletypeWriter Exchange, switched service (originally Western Union) separate from Telex.

U

u   English transliteration of Greek mu (μ), for micro or millionth; prefix in abbreviation of units like us, um.

U   Interface between CO and CP for ISDN.

U   Unnumbered format, command frames, same as UI (Layer 2).

U   Rack Unit, vertical space of 1.75 inch.

UA   Unnumbered Acknowledgement, LLC frame to accept connection request (Layer 2).

UART   Universal Async Receiver Transmitter, interface chip for serial async port.

UAS   Unavailable Second, when BER of line has exceeded 10-3 for 10 consecutive seconds until next AVS start.

UBR   Unspecified Bit Rate, service with no bandwidth reservation (ATM).

UDLC   Universal Data Link Control, Sperry Univac's HDLC

UDP/IP   Universal Data Protocol or User Datagram Protocol over Internet Protocol; UDP is a transport layer, like TCP.

uF   Microfarad, one millionth of the unit of capacitance.

UI   Unnumbered Information, frame at LLC level whose control field begins with 11: XID, TEST, SABME, UA, DM, DISC, FRMR (802).

UID	User-Interactive Data, circuit mode digital transport (ISDN).
ULP	Upper Layer Protocol.
UNI	User-Network Interface, demark point of ATM, SDH, FR, and B-ISDN at customer premises.
UNMA	Unified Network Management Architecture; AT&T's umbrella software system.
UNR	Uncontrolled Not Ready, signaling bit in CMI.
UOA	U-interface Only Activation, BRI control bit.
UP	Unnumbered Poll, command frame (Layer 2).
U-Plane	User Plane, bearer circuit for customer information, controlled by C-Plane.
UPC	Usage Parameter Control, flow control of ATM cells into network (I.555).
UPS	Uninterruptable Power Supply.
μs	Microsecond; 10^{-6} second.
USART	Universal Sync/Async Receiver Transmitter, interface chip for sync and async data I/O.
USAT	Ultra-Small Aperture SATellite; uses ground station antenna less than 1 m diameter.
USB	Universal Serial Bus.
USOC	Universal Service Order Code.
UTC	Universal Coordinated Time, the ultimate global time reference.
UTP	Unshielded Twisted Pair, copper wire used for LANs and local loops.
UUSCC	User to User Signaling with Call Control, ISDN feature that passes user data with some signaling messages.

V

V	Volt, unit of electrical potential.
V5	ETS for interface between AN and PSTN.
V.25bis	Dialing command protocol for modems, CSUs, etc.
V.35	CCITT recommendation for 48 kbit/s modem that defined a data interface; replaced by V.11 (electrical) and EIA-530 (mechanical and pinout on DB-25).
VAD	Voice Activity Detection, silence suppression, VOX.
VAN	Value Added Network; generally a packet switched network with access to data bases, protocol conversion, etc.
VBD	Voice Band Data, ISDN terminal mode that may include a modem or fax.
VBR	Variable Bit Rate, packetized bandwidth on demand, not dedicated (ATM).
VC	Virtual Container, a cell of bytes carrying a slower channel to define a path in SDH; VC-n corresponds to DS-n, n = 1 to 4.
VC	Virtual Circuit (Channel), logical connection in packet network so net can transfer data between two ports.
VCC	Virtual Circuit (Channel) Connection; between terminals (SMDS, SONET, ATM).
VCI	Virtual Circuit (Channel) Identifier; part of a packet, frame, or cell address in header (802.6, ATM).
VCL	Virtual Channel Link (ATM).
VCX	Virtual Channel Cross-connect, device to switch ATM cells on logical connections.
VDSL	Very-high-speed DSL, up to 50 Mbit/s over short loops.

VDT	Video Display Terminal, often applied to any type of "tube" or PC.
VESA	Video Electronics Standards Assoc., defined the VESA-bus for personal computers.
VF	Voice Frequency, 300-3300 Hz or up to 4000 Hz.
VFRAD	Voice FRAD, one with voice port(s) for VoFR.
VG	Voice Grade; related to the common analog phone line; 300-3000 Hz.
VGA	Video Graphics Array.
VGPL	Voice Grade Private Line, an analog line.
VHF	Very High Frequency, radio band from 30 to 300 MHz.
VLAN	Virtual LAN, term for logical LAN connectivity based on need rather than physical connection.
VLSI	Very Large Scale Integration, putting thousands of transistors on a single chip.
VMTP	Versatile Message Transport Protocol, designed at Stanford to replace TCP and TP4 in high-speed networks.
VNL	Via Net Loss, related to TLP.
VOD	Video on Demand.
VoFR	Voice Over Frame Relay.
VOX	Voice Activation, in voice over FR, silence suppression by not sending frames when audio level is below threshold.
VP	Virtual Path, for many VCCs between concentrators (ATM).
VPC	VP Connection.
VPI	VP Identifier, VCI in ETSI version of ATM; applies to bundle of VCCs between same end points (ATM).
VPT	Virtual Path Termination (ATM).
vPOTS	Very Plain Old Telephone Service; no switching; ARD etc.
VPC	Virtual Path Connection; between switches (SONET).
VPL	Virtual Path Link; between switches, may carry many connections (ATM).
VPN	Virtual Private Network, logical association of many user sites into CUG on PSTN.
VPX	Virtual Path Crossconnect; SONET device like a DACS.
VQC	Vector Quantizing Code; a voice compression technique that runs at 32 and 16 kbit/s.
VQL	Variable Quantizing Level; voice encoding method.
VR	Receive state Variable, value in register at receiver indicating next NS expected (Layer 2).
VR	Virtual Route (SNA).
VRU	Voice Response Unit, automated way to deliver information and accept DTMF inputs.
VS	Send state Variable, value in register of sender of NS in last frame sent (layer 2).
VSAT	Very Small Aperture Terminal, satellite dish under 1 m.
VSELP	Vector-Sum Excited Linear Prediction, compression algorithm used in some digital cellular systems.
VT	Virtual Tributary, logical channel made up of a sequence of cells within SONET or similar facility.
VTAM	Virtual Telecommunications Access Method, SNA protocol and host communications program.

VTE   Virtual Tributary Envelope, the real payload plus path overhead within a VT (SONET).

VTG   Virtual Trunk Group, pseudo TDM channels over ATM.

VTNS   Virtual Telecommunications Network Service.

VTOA   Voice and Telephoney Over ATM, a working group.

V.35   Former CCITT recommendation for a modem with a 48 kbit/s interface on a large 44-pin
             connector, being replaced by EIA-530 pinout on DB-25.

W

WACK   Wait before transmit positive Acknowledgment, control sequence of DLE plus second
             character (30 ASCII, 6B EBCDIC).

WAN   Wide Area Network, the T-1, T-3, or broadband backbone that covers a large
             geographical area.

WATS   Wide Area Telephone Service, large-user long distance, includes 800.

WBC   WideBand Channel, one of 16 FDDI subframes of 6.144 Mbit/s assignable to packet or
             circuit connections.

W-DCS   Wideband Digital Cross-connect System, 3/1 DACS for OC-1, STS-1, DS-3, and below,
             including T-1 (see B-DCS).

WDM   Wavelength Division Multiplexing, 2 or more colors of light on 1 fiber.

WIRE   Workable Interface Requirements Example, definition of interface between protocol lay-
             ers (ATM).

WWW   World Wide Web, information service nodes linked over the Internet.

X

X   X class central office switch is in HPR net but not linked to NP.

X   Any digit, 0-9.

X.25   CCITT recommendation defining Level 3 protocol to access a packet switched network.

xDSL   Unspecified Digital Subscriber Line method, any of various ways to send fast data on
             copper loops (ISDN, HDSL, xDSL, ADSL, etc.).

XGMON   X-Windows (based) Graphics Monitor, IBM net management software for SNMP.

XID   Exchange Identification, type of UI command to exchange parameters between LLC
             entities (layer 2).

XNS   Xerox Network Services, a LAN protocol stack.

X-off   Transmit Off, ASCII character from receiver to stop sender.

X-on   Transmit On, ok to resume sending.

XTP   eXpress Transfer Protocol, a simplified low-processing protocol proposed for broad-
             band networks.

Y   Yellow alarm control bit in sync byte (TS 24) of T1DM, Y=0 indicates alarm.

Z

Z	Impedance, nominal 600 ohm analog interface may be closer to complex value of 900 R + 2 uF C.
Z2	LOH byte used in ATM to return FEBE value.
ZBTSI	Zero Byte Time Slot Interchange, process to maintain 1's density.

Numeric

1Base5	1 Mbit/s BASEband signaling good for 500 m, STARLAN standard (802.3).
2**7	2 raised to the 7th power; exponential notation.
2B+D	Two Bearer plus a Data channel, format for ISDN basic rate access.
2B1Q	2 Binary 1 Quaternary, line code for BRI at U reference point.
2W	2-Wire, analog interface with send and receive on same pair of wires.
4B/5B	Coding that substitutes 5 bits for each 4 bits of data, leaving extra codes for commands (802.6 & FDDI). See also DMC, NRZ, AMI.
4W	4-Wire, analog or digital interface with receive and send on separate wire pairs.
5E8	Software release for 5ESS that supports National ISDN-1.
5ESS	Trademark for class 5 (end office) ESS made by AT&T.
10Base5	10 Mbit/s BASEband signaling good for 500 m, LAN definition of Ethernet (802.3).
10BaseT	10 Mbit/s BASEband signaling over twisted pair, Ethernet (802.3).
23B+D	23 Bearer Plus a Data channel, ISDN primary rate T-1 format.
30B+D	30 Bearer channels plus a Data channel, ISDN primary rate E-1 format.
800	Area code for phone service where called party pays the carrier for the call (Free Phone); also 888.
802.x	IEEE standards for LAN protocols.
802.1	Spanning tree algorithm implemented in bridges.
802.3	Ethernet.
802.4	Token Bus architecture for MAP LAN.
802.5	Token Ring, with source routing.
802.6	Distributed Queue, Dual Bus MAN.
802.10	LAN security.
802.11	Spread spectrum local radio for LANs.
900	Area code for phone service where calling party is charged for the call plus a fee that the carrier pays to called party.
54016	Specification for ESF; AT&T Pub.
62411	Basic description of T-1 service and interface; old AT&T Pub.

New Additions

You are invited to send your own list to the author for inclusion in a future edition.

Appendix B

Bibliography

Standards Library on CD-ROM
IHS Communications Products
   Twenty-two CDs of page images, and one CD index, reproduce the ANSI, ITU, ISO, EIA/TIA, and other standards as they are published by their originators. Updated regularly. A stupefying quantity of information if converted to printed pages at one time, these standards on CD are available in seconds on your PC screen.   Searches are made by key words in the descriptions as well as in the titles, which speeds you to the document you want. This set was a huge help in checking the details of signaling, data formats, and many other facts and details. Though missing specifications from some organizations who want to protect the serious money they get for paper documents, this set is well worth having, though expensive.

   Their WWWeb site is
http://www.ihscommunications.com

BOC Notes on the LEC Networks
Bellcore SA-TSV-002275
   This giant 3-ring binder (almost 8 pounds, 3.5 kg) has the scoop on what the Bell Operating Companies are actually doing in the Local Exchange Carrier arena; that is, how they deliver local phone service and interconnect to long distance companies. Of the 18 tabbed sections, the

largest is on Signaling—it covers even more types than described in Chapter 4.

Notes was published by Bellcore, which was recently purchased from the RBOCs by SAIC.

EIA/TIA-464   PBX Switching Equipment for Voiceband Applications Electronics Industry Assoc./Telecommunications Industry Assoc.

This is a design specification for PBX makers and those interested in the nittiest gritty of analog and digital voice interfaces, signaling procedures, and hardware features.

Interface Between Carriers and Customer Installations— Analog Voicegrade Switched Access Lines Using Loop Start and Ground-Start Signaling American National Standards Institute, T1.401-1993

They do try to be specific when they name documents. This American National Standard is another detailed look at analog signaling for 'Customer Installations,' also known as CPE. ANSI concentrates on the interfaces used in the U.S., a slightly different viewpoint from the equipment-oriented view of EIA-464.

Contents

Revision History

Version	Change	Date
1.0	Approved	May, 1997

1. Introduction

1.1 Purpose

Frame relay is now a major component of many network designs. The protocol provides a minimal set of switching functions to forward variable sized data payloads through a network. The basic frame relay protocol, described in the Frame Relay Forum User to Network (UNI) and Network to Network (NNI) Implementation Agreements, has been augmented by additional agreements which detail techniques for structuring application data over the basic frame relay information field. These techniques enabled successful support for data applications such as LAN bridging, IP routing, and SNA.

This specification extends frame relay application support to include the transport of digital voice payloads. Frame formats and procedures required for voice transport are described in this Implementation Agreement. This specification addresses the following requirements:

- Transport of compressed voice within the payload of a frame relay frame

- Support a diverse set of voice compression algorithms

- Effective utilization of low-bit rate frame relay connections

- Multiplexing of up to 255 sub-channels on a single frame relay DLCI

- Support of multiple voice payloads on the same or different sub-channel within a single frame

- Support of data sub-channels on a multiplexed frame relay DLCI

Transport of compressed voice is provided with a generalized frame format that supports multiplexing of sub-channels on a single frame relay DLCI. Support for the unique needs of the different voice compression algorithms is accommodated with algorithm-specific "transfer syntax" definitions. These definitions establish algorithm specific frame formats and procedures. Annexes describing different transfer syntax definitions are found at the end of this document.

Transport of supporting information for voice communication, such as signalling indications (e.g., ABCD bits), dialed digits, and facsimile data, is also provided through the use of transfer syntax definitions specific to the information being sent.

1.2 Overview of Agreement

A description of the reference model and service description for the voice over frame relay (VoFR) service is provided in Section 2, along with the concept of a Voice Frame Relay Access Device (VFRAD).

Specification of the frame formats and procedures is provided in Section 3.

Transfer syntax definitions for individual voice compression algorithms as well as generic supporting information (e.g., dialed digits) are provided in Annex sections at the conclusion of the document. Figure 1-1 illustrates some of the transfer syntax definitions used for Voice Over Frame Relay.

Page 1

Vocoders					Other			
G.729	G.728	G.723.1	G.726/G.727	G.711	Dialed	CAS	Data	Fax
CS-ACELP	LD CELP	MP-MLQ	ADPCM	PCM	Digits		Transfer	Relay

Figure 1-1 Transfer Syntax Examples

1.3 Voice Frame Relay Access Device (VFRAD)

A voice over frame relay access device supports voice services. A VFRAD may be positioned between a PBX or key set and the frame relay network. Alternatively, the VFRAD may be integrated into an end-system that directly supports telephony applications and frame relay. The VFRAD multiplexes voice and fax traffic along with data traffic from a variety of services/sources into a common frame relay connection.

1.4 Definitions

Must, Shall or Mandatory - the item is an absolute requirement of this implementation agreement.

Should - the item is desirable.

May or Optional - the item is not compulsory, and may be followed or ignored according to the needs of the implementor.

1.5 Acronyms

ADPCM	Adaptive Differential Pulse Code Modulation
AIS	Alarm Indication Signal
B_c	Committed Burst
B_e	Excess Burst
BECN	Backward Explicit Congestion Notification
BER	Bit Error Rate
CAS	Channel Associated Signalling
CS-ACELP	Conjugate Structure – Algebraic Code Excited Linear Predictive
CELP	Code Excited Linear Prediction
CID	Channel Identification
CIR	Committed Information Rate
CCS	Common Channel Signalling
DE	Discard Eligibility

DLCI	Data Link Connection Identifier
DTMF	Dual Tone Multi-Frequency
E-ADPCM	Embedded Adaptive Differential Pulse Code Modulation
FAX	Facsimile Group 3
FECN	Forward Explicit Congestion Notification
FRAD	Frame Relay Access Device
HDLC	High Level Data Link Control
IA	Implementation Agreement
I/F	Interface
IWF	Inter-working Function
LD-CELP	Low Delay - Code Excited Linear Prediction
lsb	Least Significant Bit
MP-MLQ	Multi Pulse Maximum Likelihood Quantizer
msb	Most Significant Bit
PCM	Pulse Code Modulation
PVC	Permanent Virtual Connection
SID	Silence Information Descriptor
UNI	User Network Interface
VAD	Voice Activity Detection
VFRAD	Voice Frame Relay Access Device
VoFR	Voice Over Frame Relay
Vocoder	Voice coder/decoder

1.6 Relevant Standards

[1]　　FRF.1.1　　　Frame Relay User-to-Network Implementation Agreement, January 1996

[2]　　FRF.3.1　　　Multiprotocol Encapsulation Implementation Agreement, June 22, 1995

[3]　　FRF.12　　　Frame Relay Fragmentation Implementation Agreement, 1997

[4]　　ITU G.711　　Pulse Code Modulation of Voice Frequencies, 1988

[5]　　ITU G.723.1　Dual Rate Speech Coder for Multimedia Communications Transmitting at 5.3 & 6.3 kbit/s, March 1996

[6]　　ITU G.726　　40, 32, 24, 16 kbit/s Adaptive Differential Pulse Code Modulation (ADPCM), March 1991

[7]　　ITU G.727　　5-,4-,3- and 2 bits Sample Embedded Adaptive Differential Pulse Code Modulation, November 1994

[8]　　ITU G.728　　Coding of Speech at 16 kbit/s Using Low-Delay Code Excited Linear Prediction, November 1994

[9]　　ITU G.729　　Coding of Speech at 8kbit/s using Conjugate Structure - Algebraic Code Excited Linear Predictive (CS-ACELP) Coding, March 1996

[10]　ITU G.764　　Voice Packetization – Packetized voice protocols, December 1990

[11]　ITU T.30　　　Terminal Equipment and protocol for Telematic Service/Procedure for Facsimile General Switch Networks, March 1993

2. Reference Model and Service Description

2.1 Frame Relay Access

A VFRAD uses the frame relay service at the UNI as a transmission facility for voice, voice signalling, and data. The reference model for voice over frame relay is shown in Figure 2-1. Using the Voice over Frame Relay (VoFR) service, it is possible for any type of VFRAD on the left-hand side of Figure 2-1 to exchange voice and signalling with any type of VFRAD on the right-hand side of Figure 2-1.

Three types of devices are shown in Figure 2-1. The top layer shows end-system devices similar to telephones or FAX machines. The middle layer shows transparent multiplexing devices similar to channel banks. The bottom layer shows switching system devices similar to PBX's.

A VFRAD connects to a frame relay UNI via physical interfaces as defined in [1].

2.1.1 End-System Devices

The top left device in Figure 2-1 could be a PC with FAX or telephony application software using a frame relay network port for connectivity to other VFRAD devices. Such an end-system could use the VoFR protocol stack on a frame relay connection to another end-system (top right). It could also use the VoFR protocol stack on a connection to a transparent channel bank into a private network (middle right) or to a PBX (bottom right).

2.1.2 Transparent-Multiplexing Devices

The middle left device in Figure 2-1 could be a Channel Bank connected via analog trunks to an external PBX (not shown). Such a multiplexing device could use the VoFR protocol stack on a frame relay connection to an end-system (top right), another channel bank (middle right) or a PBX (bottom right).

2.1.3 Switching-System Devices

The bottom left device in Figure 2-1 could be a PBX using a frame relay network for connection to off premise extensions (end-systems) or as trunks to other PBX devices. Such a switching system device could use the VoFR protocol stack on a frame relay connection to an end-system (top right), a channel bank (middle right), or another PBX (bottom right).

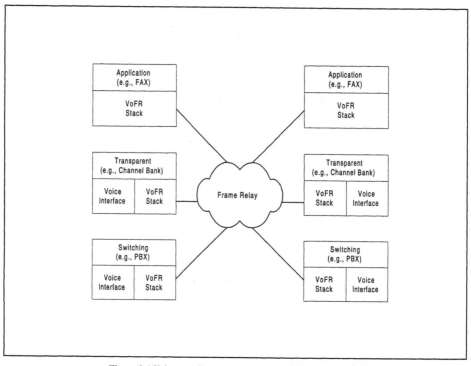

Figure 2-1 Voice over Frame Relay Network Reference Model

2.2 Voice Interfaces

The requirements for implementation of voice interfaces are beyond the scope of this implementation agreement.

2.3 Voice Over Frame Relay Service Description

This implementation agreement defines formats and procedures that support a VoFR service. Elements of the VoFR service support various types of service users that may be performing any of the following voice applications:

1. Call origination and termination — for an end-system

2. Transparent interworking between individual sub-channels on a VoFR interface and sub-channels on another type of voice interface.

3. Call-by-call Switching — for a switching system to terminate an incoming call and originate a call on another voice interface.

To support the VoFR service, the underlying protocol stack must provide a full duplex transport service. The service users can use the following service elements to operate a voice connection. The service elements support the transport of two types of payloads: primary payloads and signalled payloads. Refer to Section 3.1 for a discussion of payload types.

2.3.1 Primary Payloads

2.3.1.1 Encoded Voice

This service element conveys voice information supplied by the service user. The voice information is packaged according to the rules specified by a voice transfer syntax. Voice transfer syntax definitions for various voice compression schemes are described in the annexes of this IA.

2.3.1.2 Encoded FAX or Voice-Band Modem Data

The service users can exchange digital data in a "baseband" format suitable for re-modulation into a FAX or analog modem signal. The VoFR service transports this information between the two service users.

The transmitting service user may locally detect the presence of a FAX or voice-band modem signal for the voice connection and demodulate it before sending it. The receiving service user can detect arriving packets that contain demodulated data and can reconstruct the original modulated signal instead of reconstructing a speech signal.

The encoded FAX or voice-band data payload format is within the scope of this IA. The algorithms used for demodulation and re-modulation of FAX and/or voice-band data are outside the scope of this IA.

The transfer syntax for FAX is described in Annex D.

The transfer syntax for voice band modem data is for further study.

2.3.1.3 Data Frames

This service element conveys data frames supplied by the service user. The frames are packaged according to the rules specified by Annex C.

The content of the data frames is transparent to the VoFR service.

One application of the data frame service element enables transparent tunneling of common channel signalling messages between two compatible end-points (e.g., PBX interfaces). Common channel signalling message formats and procedures are beyond the scope of this agreement.

2.3.2 Signalled Payload

2.3.2.1 Dialed Digits

This service element transparently conveys DTMF, pulse, or other dialed digits supplied by the service user. These digits may be sent during the voice call setup or following call establishment to transfer in-band tones.

2.3.2.2 Signalling Bits (Channel Associated Signalling)

This service element transparently conveys signalling bits supplied by the service user. These bits may indicate seizure and release of a connection, dial pulses, ringing, or other information in accordance with the signalling system in use over the transmission facility.

2.3.2.3 Fault Indication

The service users can use this service to convey an alarm indication signal.

2.3.2.4 Message-Oriented Signalling (Common Channel Signalling)

Refer to section 2.3.1.3

2.3.2.5 Encoded FAX

Refer to Section 2.3.1.2

Encoded FAX may be transmitted on a sub-channel that utilizes a primary payload for encoded voice. In this case, the sub-frames containing the encoded FAX must be sent as a signalled payload.

2.3.2.6 Silence Information Descriptor

Silence Information Descriptor (SID) sub-frames indicate the end of a talk-spurt and convey comfort noise generation parameters. These SID indications support voice activity detection (VAD) and silence suppression schemes.

When VAD is utilized, a SID sub-frame may optionally be transmitted following the last encoded voice sub-frame of a talk-spurt. Reception of a SID sub-frame after a voice sub-frame may be interpreted as an explicit indication of end of talk-spurt. In addition, SID sub-frames may be transmitted at any time during the silence interval to update comfort noise generation parameters.

The SID payload is defined for PCM and ADPCM encoding in the appropriate annexes (Annex F and Annex G). The SID payload definition for other voice encoding algorithms is for further study and can be null.

SID sub-frames should not be sent if VAD is not utilized.

2.4 VFRAD Configuration Requirements

VoFR devices compliant with this implementation agreement are not required to negotiate operational parameters. Negotiation procedures are for further study. Therefore, at the time of provisioning, the network manager must configure end-to-end configuration parameters (e.g., Vocoder.). End-point devices providing the VoFR service are configured with compatible sub-channel assignments, signalling, compression algorithms, and other options.

2.5 VoFR Service Block Diagram

The relationship of the Voice Over Frame Relay service, VoFR Service user and the frame relay service is shown in Figure 2-2.

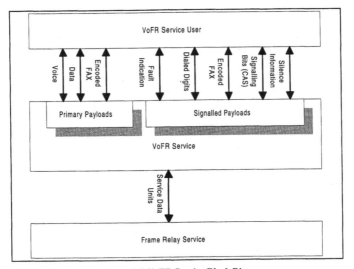

Figure 2-2 VoFR Service Block Diagram.

2.6 Service Multiplexing

The Frame Relay UNI can support multiple PVCs, each of which can provide VoFR service. The VoFR service supports multiple voice and data channels on a single frame relay data link connection. The VoFR service delivers frames on each sub-channel in the order they were sent.

As shown in Figure 2-3 each instance of the voice/data multiplexing layer can support one or more voice connections and data protocol stacks over a single frame relay PVC. The mechanism for separation of the voice and data connections being supported over a single frame relay PVC is within the scope of this IA. The mechanisms and protocol stacks used for data connections are covered in other Frame Relay Forum IA's and relevant standards.

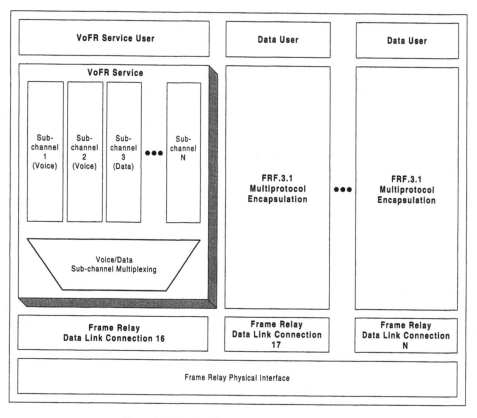

Figure 2-3 Voice Over Frame Relay Multiplexing Model

3. Frame Formats

Voice and data payloads are multiplexed within a voice over frame relay data link connection by encapsulation within the frame format specified in [1]. Each payload is packaged as a sub-frame within a frame's information field. Sub-frames may be combined within a single frame to increase processing and transport efficiencies. Each sub-frame contains a header and payload. The sub-frame header identifies the voice/data sub-channel and, when required, payload type and length. Refer to Figure 3-1 for an illustration of sub-frames. In this example, a single DLCI supports 3 voice channels and 1 data channel. Three voice payloads are packaged in the first frame and a data payload is contained in the second frame.

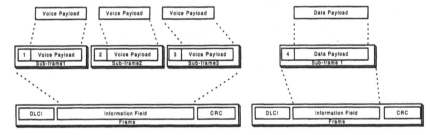

Figure 3-1 Relationship between frames and sub-frames

3.1 Payloads

3.1.1 Primary payload

Each sub-channel of a VoFR connection transports a primary payload. A primary payload contains traffic that is fundamental to operation of a sub-channel. Other payloads may be sent to support the primary payload (e.g., dialed digits for a primary payload of encoded voice). These additional payload types are differentiated from the primary payload by a signalled encoding in the payload type field of the sub-frame. A payload type of all zeros always indicates the primary payload.

Three basic types of primary payloads are utilized: encoded voice payloads, encoded FAX payloads, and data payloads. Refer to the appropriate annex for a description of the transfer syntax which supports these payload types.

3.1.2 Signalled Payload

Payloads containing in-band information, which augment the primary payload flow, are indicated using payload type codings. These signalled payloads include information such as channel-associated signalling, dialed digits, in-band encoded FAX relay, and fault indications. Refer to the appropriate annex for a description of the service elements which support the signalled payloads.

3.2 Sub-frame Format

Each sub-frame consists of a variable length header and a payload. The minimal sub-frame header is a single octet containing the least significant bits of the voice/data channel identification along with extension and length indications. An extension octet containing the most significant bits of the voice/data channel identification and a payload type is present when the Extension Indication is set. A payload length octet is present when the Length Indication is set. Refer to Figure 3-2 and Table 3-1 for a description of the sub-frame structure.

Bits

8	7	6	5	4	3	2	1	Octets
EI	LI	Sub-channel Identification (CID) (Least significant 6 bits)						1
CID (msb)	0 Spare	0 Spare	Payload Type					1a (Note 1)
Payload Length								1b (Note 2)
Payload								n

NOTES:
1. When the EI bit is set, the structure of Octet 1a given in Table 3-1 applies.
2. When the LI bit is set, the structure of Octet 1b given in Table 3-1 applies.
3. When both the EI bit and the LI bit are set to 1 both Octet 1a and 1b are used.

Figure 3-2 Sub-frame format

Extension indication (octet 1)

The extension indication (EI) bit is set to indicate the presence of octet 1a. This bit must be set when a sub-channel identification value is > 63 or when a payload type is indicated. Each transfer syntax has an implicit payload type of zero when the EI bit is cleared.

Length indication (octet 1)

The length indication (LI) bit is set to indicate the presence of octet 1b. The LI bit of the last sub-frame contained within a frame is always cleared and the payload length field is not present. The LI bits are set for each of the sub-frames preceding the last sub-frame.

Sub-channel identification (octets 1 and 1a)

The six least significant bits of the sub-channel identification are encoded in octet 1. The two most significant bits of the sub-channel identification are encoded in octet 1a. A zero value in the two most significant bits is implied when octet 1a is not included in the VoFR header (EI bit cleared). Sub-channel identifiers 0000 0000 through 0000 0011 are reserved in both the short and long format.

Payload type (octet 1a)

This field indicates the type of payload contained in the sub-frame.

Bits				
4	3	2	1	
0	0	0	0	Primary payload transfer syntax
0	0	0	1	Dialed digit transfer syntax (Annex A)
0	0	1	0	Signalling bit transfer syntax (Annex B)
0	0	1	1	Fax relay transfer syntax (Annex D)
0	1	0	0	Silence Information Descriptor

A zero value for the payload type is implied when octet 1a is not in included in the header (EI bit cleared).

Payload length (octet 1b)

Payload length contains the number of payload octets following the header. A payload length indicates the presence of two or more sub-frames packed in the information field of the frame.

Payload (octet p)

The payload contains octets as defined by the applicable transfer syntax assigned to the sub-channel or as indicated by the payload type octet 1a.

Table 3-1 Sub-frame format

3.3 Sub-frame Examples

The diagrams in this section illustrate some of the possible combinations of sub-frames. Figure 3-3 shows a frame which contains a single voice payload for a low-numbered sub-channel. Octets 1a and 1b are not required. The payload, a CS-ACELP sample, starts after octet 1.

Figure 3-3 Frame containing one sub-frame

Figure 3-4 shows a frame which contains a single voice payload for a high-numbered channel (>63). Octet 1a must be included. Note that the payload type is zero, indicating the transfer syntax that has been configured for the channel. In this example, the transfer syntax is the CS-ACELP syntax.

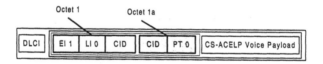

Figure 3-4 Frame containing one subframe for a high numbered channel

Figure 3-5 shows a frame which contains multiple sub-frames for channels 5 and 6. In this case, the payload type is non-zero and octet 1a is required to encode the payload type. The first of the two sub-frames includes octet 1b with the encoding of payload length.

Figure 3-5 Frame containing multiple sub-frames

Figure 3-6 shows a frame which contains multiple sub-frames for channels 5 and 6. In this case, the payload type is zero and the payload length (octet 1b) appears in the first of the two sub-frames.

Figure 3-6 Frame containing multiple sub-frames

4. Minimum Requirements for Conformance

This agreement provides support for several optional transfer syntax definitions. Interoperability between VoFR devices is possible only when both devices share support for one or more common transfer syntax definitions. VoFR devices are classified based on the support provided for the common transfer syntax definitions. Class 1 compliant devices support capabilities suitable for high bit-rate interfaces. Class 2 compliant devices support capabilities that enable optimal performance over low bit-rate frame relay interfaces. An implementation is compliant with this agreement if the requirements for at least one of the two classes are met.

4.1 Class 1 Compliance Requirements

4.1.1 Frame Formats

1. Support the frame structure described in section 3.

2. Received optional frames may be discarded.

4.1.2 Primary Payload Types

1. Support of G.727 as described in Annex F is mandatory. Support of other vocoders described in Annex F is optional.

2. A transmit rate of 32Kbps is mandatory.

3. Support for rates of 32kbps, 24kbps, 16kbps are mandatory at the receiver.

4. Support for other primary payload transfer syntax definitions (e.g., FAX) is optional.

4.1.3 Signalled Payload Types

1. Support for the dialed digit signalled payload type is optional.

2. Support for the signalling bits signalled payload type (CAS and AIS) is mandatory.

3. Support for the encoded FAX signalled payload type is optional.

4.2 Class 2 Compliance Requirements

4.2.1 Frame Formats

1. Support the frame structure described in section 3.

2. Received optional frames may be discarded.

4.2.2 Primary Payload Types

1. Support for Annex E CS-ACELP G.729 or G.729A voice transfer syntax is mandatory.

5. Support for other primary payload transfer syntax definitions (e.g., FAX) is optional.

4.2.3 Signalled Payload Types

1. Support for the dialed digit signalled payload type is mandatory.

2. Support for the signalling bits signalled payload type (CAS and AIS) is mandatory.

3. Support for the encoded FAX signalled payload type is optional.

Annex A – Dialed Digit Transfer Syntax

A.1   Reference Documents

None

A.2   Transfer Structure

The Dialed Digit Transfer Syntax is comprised of the Dialed Digit Payload Format and the Dialed Digit Transfer Procedure.

A.3   Dialed Digit Payload Format

At the originating VFRAD the detected digits are inserted into a Dialed Digit Payload by the Dialed Digits Service element. Payload carrying digits will be identified using the VoFR sub-frame payload type field codepoint for the Dialed Digit transfer syntax. The digits will automatically be associated with the corresponding voice traffic based on the Channel ID field.

Each Digit Payload contains three windows of digit transition. The first window represents the current 20ms period [0], the second [-1] is the recent period and the third [-2] is the previous.

Bits

8	7	6	5	4	3	2	1	Octet
			Sequence Number					P
reserved 000			Signal Level					P+1
Digit Type [0]			Edge Location [0]					P+2
reserved 000			Digit Code [0]					P+3
Digit-Type[-1]			Edge-Location[-1]					P+4
reserved 000			Digit-Code[-1]					P+5
Digit-Type[-2]			Edge-Location[-2]					P+6
reserved 000			Digit-Code[-2]					P+7

Figure A - 1 Dialed Digit Payload Type

A.3.1   Sequence Number

The sequence field is an 8-bit number that is incremented for every fragment transmitted. The sequence field wraps from all ones to zero in the usual manner of such sequence numbers. Each increment of the sequence represents a period of 20ms.

A.3.2   Signal Level

The power level of each frequency is between 0 to -31 in -dBm0. Power levels above zero dBm0 are coded 00000. In the event that one dialed digit payload contains a transition from one dialed digit to another dialed digit, the signal level field applies to the dialed digit in the "current" 20 ms period.

Code	Power Level dBm0
00000	0
00001	-1
000010	-2
000011	-3
00100	-4
00101	-5
00110	-6
00111	-7
01000	-8
01001	-9
01010	-10
01011	-11
01100	-12
01101	-13
01110	-14
01111	-15
10000	-16
10001	-17
10010	-18
10011	-19
10100	-20
10101	-21
10110	-22
10111	-23
11000	-24
11001	-25
11010	-26
11011	-27
11100	-28
11101	-29
11110	-30
11111	-31

Figure A - 2 Signal Level

A.3.3   Digit Type

Code	Digit Type
000	Digit Off
001	DTMF On
010-111	Reserved

Figure A - 3 Digit Types

A 20ms window is used to encode the edge when a digit is turned on and off. This is the delta time, 0ms (00000) to 19ms (10011), from the beginning of the current frame in ms. If there is no transition, the edge location will be set to 0 and the Digit Type of the previous windows will be repeated.

A.3.4   Digit-Code

The following DTMF digit codes are encoded when dialed digit type = DTMF ON.

Digit Code	DTMF Digits
00000	0
00001	1
00010	2
00011	3
00100	4
00101	5
00110	6
00111	7
01000	8
01001	9
01010	*
01011	#
01100	A
01101	B
01110	C
01111	D
10000-11111	Reserved

Figure A - 4 DTMF Digit Codes

A.4   Dialed Digit Transfer Procedures

A.4.1   Procedure for Transmission of Dialed Digit Payloads

When the transmitter detects a validated digit, or has addressing information to send it will start sending a Dialed Digit Payload every 20ms. Since each payload covers 60ms of Digit on/off edge information, there is redundancy of the edge information. The sequence number is incremented by one in each transmitted payload.

When the digit activity is off, the transmitter should continue to send three more Dialed Digit Payloads for 60ms.

A.4.2   Procedure for Interpreting Received Dialed Digit Payloads

When the receiver gets a Dialed Digit Payload or accepts the received addressing information, it will generate digits according to the location of the on and off edges. Silence will be applied to the duration after an off edge and before an on edge. Digits will be generated after an on edge and before an off edge.

If the sequence number is one greater than the last received sequence number, the receiver appends the Current edge information to the previously received information.

If the sequence number is two greater than the last received sequence number, the receiver appends the Recent and Current edge information to the previously received information.

If the sequence number is three greater than the last received sequence number, the receiver appends the previous, recent and current edge information to the previously received information.

If the sequence number is more than three greater than the last received sequence number, the receiver appends the previous, recent and current edge information to the previously received information. It fills in the gap with the static values based on the previously received payload.

On a given sub-channel, if a voice payload is received at any time, an off edge should be appended to the previously received digits on/off edge information.

Annex B – Signalling Bit Transfer Syntax

B.1   Reference Documents

None

B.2   Transfer Structure

The Signalling Bit Transfer Syntax is comprised of payload formats and transfer procedures for alarm indications and channel associated signalling bits.

B.3   Payload Format

Payloads carrying signalling bits will be identified using the payload type field in the VoFR Header. The signalling bits will automatically be associated with the corresponding voice traffic based on the Channel ID field.

The first byte following the VoFR header contains a seven-bit sequence number with the most significant bit assigned as an Alarm Indicator Signal (AIS) bit. A value of 1 signifies an alarm condition.

Bits

	8	7	6	5	4	3	2	1	Octet
	AIS	Sequence Number							P
Previous	D[t-56ms]	C[t-56ms]	B[t-56ms]	A[t-56ms]	D[t-58ms]	C[t-58ms]	B[t-58ms]	A[t-58ms]	P+1
	D[t-52ms]	C[t-52ms]	B[t-52ms]	A[t-52ms]	D[t-54ms]	C[t-54ms]	B[t-54ms]	A[t-54ms]	P+2
	D[t-48ms]	C[t-48ms]	B[t-48ms]	A[t-48ms]	D[t-50ms]	C[t-50ms]	B[t-50ms]	A[t-50ms]	P+3
	D[t-44ms]	C[t-44ms]	B[t-44ms]	A[t-44ms]	D[t-46ms]	C[t-46ms]	B[t-46ms]	A[t-46ms]	P+4
	D[t-40ms]	C[t-40ms]	B[t-40ms]	A[t-40ms]	D[t-42ms]	C[t-42ms]	B[t-42ms]	A[t-42ms]	P+5
Recent	D[t-36ms]	C[t-36ms]	B[t-36ms]	A[t-36ms]	D[t-38ms]	C[t-38ms]	B[t-38ms]	A[t-38ms]	P+6
	D[t-32ms]	C[t-32ms]	B[t-32ms]	A[t-32ms]	D[t-34ms]	C[t-34ms]	B[t-34ms]	A[t-34ms]	P+7
	D[t-28ms]	C[t-28ms]	B[t-28ms]	A[t-28ms]	D[t-30ms]	C[t-30ms]	B[t-30ms]	A[t-30ms]	P+8
	D[t-24ms]	C[t-24ms]	B[t-24ms]	A[t-24ms]	D[t-26ms]	C[t-26ms]	B[t-26ms]	A[t-26ms]	P+9
	D[t-20ms]	C[t-20ms]	B[t-20ms]	A[t-20ms]	D[t-22ms]	C[t-22ms]	B[t-22ms]	A[t-22ms]	P+10
Current	D[t-16ms]	C[t-16ms]	B[t-16ms]	A[t-16ms]	D[t-18ms]	C[t-18ms]	B[t-18ms]	A[t-18ms]	P+11
	D[t-12ms]	C[t-12ms]	B[t-12ms]	A[t-12ms]	D[t-14ms]	C[t-14ms]	B[t-14ms]	A[t-14ms]	P+12
	D[t-8ms]	C[t-8ms]	B[t-8ms]	A[t-8ms]	D[t-10ms]	C[t-10ms]	B[t-10ms]	A[t-10ms]	P+13
	D[t-4ms]	C[t-4ms]	B[t-4ms]	A[t-4ms]	D[t-6ms]	C[t-6ms]	B[t-6ms]	A[t-6ms]	P+14
	D[t]	C[t]	B[t]	A[t]	D[t-2ms]	C[t-2ms]	B[t-2ms]	A[t-2ms]	P+15

Figure B - 1 Signalling Bit Transfer Syntax Payload Format

The sequence number starts at 0 and increments by 1 up through 127 and rolls over back to 0.

The transfer syntax for signalling bits contains 60 milliseconds worth of samples for up to four signalling bits. Each sample has a time resolution of 2.0 milliseconds. Each payload contains ten "new" samples for the current 20 millisecond time interval and a repetition of the ten samples for each of the two immediately preceding 20 millisecond time intervals.

This will result in 15 bytes of packed signalling bit values.

For sixteen state coding, all four bits are independent. For four state coding, the A and B bits are repeated in the C and D bit fields respectively. For two state coding, the A bit is repeated in the B, C, and D bit fields.

B.4   Procedures for Transmission of Payloads

While there are transitions occurring in the signalling bit values, the transmitter sends a signalling bit payload every 20 milliseconds. Since each payload covers 60 milliseconds of signal bit states, there is redundancy of signal bit information. The sequence number is incremented by one in each transmitted payload.

When the signal bit values have been static for 500 milliseconds, the transmitter switches frequency of transmission and sends a signal bit payload only once in every 5 seconds. During this time, the sequence number is not incremented.

When transitions start occurring again, the transmitter resumes incrementing the sequence numbers by one and sending payloads every 20 milliseconds.

The first such payload contains ten static previous and ten static recent values with ten new current samples. The second such payload contains ten static previous values with the ten previous values that were current in the first payload, and ten new current samples. This restarts the overlapping redundancy of information.

The transmitter may debounce the sequence of signalling bit values prior to transmission, but is not required to do so.

B.5   Procedures for Interpreting Received Payloads

When the receiver gets a signalling bit payload, it processes the bits based on the sequence number.

If the sequence number is one larger than the last received sequence number, the receiver appends the Current signal bits to the previously received values.

If the sequence number is two larger than the last received sequence number, the receiver appends the Recent and Current signal bits to the previously received values.

If the sequence number is three larger than the last received sequence number, the receiver appends the Previous, Recent, and Current signal bits to the previously received values.

If the sequence number is more than three larger than the last received sequence number, the receiver appends the Previous, Recent, and Current signal bits to the previously received values. It fills in the gap with static values based on the previously received payload.

If the sequence number is the same as the last received sequence number, the receiver takes the first value and uses it to set its current values for the signalling bits. (The signal bit values are static.)

The transmitter may or may not have debounced the signal bit values before transmission. If the receiving VoFR service user is interpreting the semantics of the signal bits, it should debounce the sequence of bit values received.

Annex C – Data Transfer Syntax

C.1   Reference Documents

[1]   FRF.12   Frame Relay Forum Fragmentation Implemention Agreement, March 1997
[2]   FRF3.1   Multiprotocol Encapsulation Implementation Agreement, June 1995
[3]   RFC 1490   Multiprotocol Interconnect over Frame Relay, 1993

C.2   Data Transfer Structure

This annex describes a transfer syntax to support transport of data frames between two voice over frame relay service users. The contents of the frames are transparent to the voice over frame relay service. Typical applications include the transport of common channel signalling messages, RFC1490 packets [3], and FRF3.1 packets [2].

All data sub-frames contain the fragmentation header.

The payload type is set to primary payload type.

For more information on the fragmentation procedure refer to [1].

C.3   Data Payload Format

Figure C-1 shows the sub-frame payload format.

			Bits					
8	7	6	5	4	3	2	1	Octet
VoFR Sub-frame Header								1
B	E	0	Sequence number (upper 5 bits)					P
Sequence number (lower 8 bits)								P+1
Payload Fragment (variable length)								P+2
								P+N

Figure C - 1 Data Transfer Syntax Payload Format

C.4   Data Procedures

The sub-frame payload will consist of a data frame received from a VoFR service user.

The frame is transmitted on the data link connection in one or more data fragments as defined in [1].

Upon receipt of a sub-frame containing the data transfer syntax, the fragments are re-combined using the procedures of [1], and the frame is delivered to the VoFR service user.

All frames received from the VoFR service user are conveyed without interpretation. Information transmitted using this transfer syntax is transparent to the VoFR service.

The maximum fragment size is governed by the maximum frame size supported by the Q.922 data link connection.

Annex D – Fax Relay Transfer Syntax

D.1 Reference Documents

[1] ITU T.4   Standardization of group 3 facsimile apparatus for document transmission, March 1993

[2] ITU T.30   Terminal Equipment and Protocol for Telematic Service / Procedure for Facsimile General Switch Networks, November 1994

[3] ITU V.17   A 2-wire modem for facsimile applications with rates up to 14,400bit/s, February 1991

[4] ITUV.21   300 bit/s duplex modem standardized for use in general switched telephone network Blue Book Fasc. VIII.1, October 1994

[5] ITU V.27   4800/2400 bit/s modem standardized for use in general switched telephone network Blue Book Fasc. VIII.1, November 1994

[6] ITU V.29   9600 bit/s modem standardized for use in point-to-point 4-wire leased telephone-type circuits Blue Book Fasc. VIII.1, November 1988

[7] ITU V.33   14400 bit/s modem standardized for use in point-to-point 4-wire leased telephone-type circuits Blue Book Fasc. VIII.1, November 1988

D.2 FAX Transfer Structure

The Fax Relay Transfer Syntax is comprised of the Fax Relay Payload Format and the Fax Relay Transfer Procedure. The fax relay transfer syntax provides transfer syntax for fax.

D.3 Fax Relay Payload Format

D.3.1 Modulation Turn-On Payload

The Modulation Turn-On Payload has the modulation types defined in the octet following the time stamp. If Modulation Type is single Frequency Tone, Frequency MS:LS bytes will specify the frequency. Frequency MS:LS bytes should be set to zero if the Modulation Type is not Single Frequency Tone.

Bits

8	7	6	5	4	3	2	1	Octet
EI1=1	Sequence Number				Relay Command=001			P
Time Stamp LS byte								P+1
EI2=0	Time Stamp MS byte							P+2
HDLC	reserved				Modulation Type			P+3
Frequency LS byte								P+4
Frequency MS byte								P+5

*LS=least significant, MS=most significant

Figure D - 1 Modulation Turn-On Payload

D.3.2   Modulation Turn-Off Payload

The Modulation Turn-Off Payload has the following structure.

Bits

8	7	6	5	4	3	2	1	Octet
EI1=1	Sequence Number				Relay Command=000			P
Time Stamp LS byte								P+1
EI2=0	Time Stamp MS byte							P+2

Figure D - 2 Modulation Turn-Off Payload

D.3.3   T.30 Payload

The T.30 Payload has 3 bytes of demodulated and HDLC de-framed data.

Bits

8	7	6	5	4	3	2	1	Octet
EI1=0	Sequence Number				Relay Command = 010,011or 100			P
Data [I]								P+1
Data[I-1]								P+2
Data[I-2]								P+3

Figure D - 3 T.30 Payload

D.3.4   T.4 Payload

The T.4 Payload should be sent once every 40ms. The Relay Command = 010 (Data). Use of the Relay Command 011 and 100 for T.4 Payload is for further study.

Bits

8	7	6	5	4	3	2	1	Octet
EI1=0	Sequence Number				Relay Command = 010			P
Data[I]								P+1
Data[I-1]								P+2
•								•
Data[I-N-1]								P+N

Figure D - 4 T.4 Payload

Each payload should have the following number of raw demodulated data bytes according to the modulation rate:

Modulation Rate	Bytes per Payload (N)
14400	72
12000	60
9600	48
7200	36
4800	24
2400	12

Figure D - 5 Modulation Rates

D.3.5   Relay Command

The following is a list of code-points for Relay Commands.

Code	Relay Command
000	Modulation-Off
001	Modulation-On
010	Data
011	HDLC-End-Of-Frame
100	HDLC-Frame-Abort
101 - 111	reserved

Figure D - 6 Relay Commands

D.3.6   Modulation Type

The following is an encoding table for the Modulation Type. However, it is optional to support all the types.

Code	Modulation Type
0000	Single Freq Tone
0001	V.21 300bps
0010	V.27ter 2400bps
0011	V.27ter 4800bps
0100	V.29 7200bps
0101	V.29 9600bps
0110	V.33 12000bps
0111	V.33 14400bps
1000	V.17 7200bps
1001	V.17 9600bps
1010	V.17 12000bps
1011	V.17 14400bps
1100 - 1111	reserved

Figure D - 7 Modulation Type

D.3.7   HDLC

HDLC=1 indicates that HDLC de-framing is being applied. HDLC=0 indicates that HDLC de-framing is not used. De-framed packets are the content that remains after removing flags, extra bits inserted for transparency, and the frame check sequence.

D.3.8   Sequence Number

This sequence number is incremented for each new packet independent of the value of Relay Command.

The sequence number is reset at the beginning of each modulation type when the Modulation-On Relay Command is sent. It wraps around when it reaches a maximum count.

D.3.9   Time Stamp

The Time Stamp information represents the relative timing of events on the analog (or equivalent) input to the demodulator. The unit for Time Stamp is 1ms. The accuracy of Time Stamp should be within +/- 5ms.

The Time Stamp is mandatory in the packet header when the Modulation-On or Modulation-Off Relay Commands are sent. It is optional with any other Relay Command.

The Time Stamp clocks free-run on each end and there is no synchronization between them. The Time Stamp wraps around when it reaches the maximum count.

D.3.10   EI1 and EI2

These are header Extension Indicator bits.

EI1=1 indicates that the two Time Stamp bytes exist and immediately follow the first header byte. EI1=0 indicates that there is no Time Stamp bytes.

EI2 is reserved for future use and should be set to 0.

D.3.11   Frequency LS & MS Bytes

These are the least significant (LS) and most significant (MS) bytes for the Single Frequency Tone in unit of Hertz (Hz) within +/- 1.5%.

D.3.12   Data

Data is packed into the packet with the latest byte first and the oldest byte last. Within each byte, the MSB is the most recent bit and LSB is the oldest bit.

D.4   Fax Relay Transfer Procedures

D.4.1   Procedure for Transmission of T.30 Data

When the preamble is detected, at least three identical Modulation Turn-On payloads should be sent with Relay Command = 001 (Modulation-On), Sequence Number = 0, EI1=1, Modulation Type = V.21 and HDLC=1. The same Time Stamp should be use in all three payloads.

When the first byte of HDLC data is being demodulated and deframed, it should be sent with Relay Command = 010 (Data) and Sequence Number =1. Data [I] is the first byte. Data [I-1] and Data [I-2] should be set to all 1's.

When the second byte of HDLC data is being demodulated and deframed, it should be sent with Relay Command = 010 (Data) and Sequence Number =2. Data [I] is the second byte, Data [I-1] is the first byte and Data [I-2] is set to all 1's.

When the third byte of HDLC data is being demodulated and deframed, it should be sent with Relay Command = 010 (Data) and Sequence Number =3. Data [I] is the third byte, Data [I-1] is the second byte and Data [I-2] is the first byte.

Subsequently, a new payload should be sent after every byte being demodulated. The payload is sent with Relay Command = 010 (Data) and the Sequence Number incremented by 1. The most current byte should be immediately after the header, followed by the recent byte and then the previous byte.

At the end of each HDLC frame, if there is no CRC error, the last payload should be sent three times with Relay Command=011 (HDLC-End-Of-Frame). If a CRC error was detected by the sender, the last payload should be sent three times with Command = 100 (HDLC-Frame-Abort). In both cases, all three payloads should have the same three bytes of data as the previous data payload. All three payloads should have the sequence number (N_{EOF}). The previous data payload should have sequence number $N_{EOF} -1$.

The first data payload of the following HDLC frame should have sequence number $N_{EOF} +1$. The data bytes Data [I-1] and Data [I-2} in the first data payload of the following frame are the last two data bytes from the previous frame.

If the modulation turns off, three identical payloads should be sent with Command = 000 (Modulation-Off). All three payloads should have the same sequence number, which is one more than the last data payload, EI=1 and the same two bytes of Time Stamp.

D.4.2   Procedure for Transmission of T.4 Data

When modulation (non-single frequency tone) for the T.4 procedure is detected, at least three Modulation Turn-On payloads should be sent with Relay Command = 001 (Modulation-On), Sequence Number = 0, EI1=1, Modulation Type = the codepoint of the detected modulation type and HDLC=0. Use of HDLC=1 is for further study. The same Time Stamp should be use in all three payloads.

Subsequently, when data is available, a payload should be sent every 40ms with relay Command = 010 (Data) and the sequence number should be incremented by 1. Use of Relay Command 011 and 100 is reserved for further study.

When the modulation turns off, three identical payloads should be sent with Command = 000 (Modulation-Off) and EI1=1. These last three payloads should have the same sequence number and the same Time Stamp. The sequence number should be one larger than the last data.

D.4.3   Handling of Non-Standard Facilities (NSF) Frame

Procedures for disabling the NSF frame are for further study.

ANNEX E – CS-ACELP Transfer Syntax

E.1 Reference Document

[1] ITU G.729/ Coding of Speech at 8 kbit/s using Conjugate Structure-Algebraic Code
     ITU G.729   Excited Linear Predictive (CS-ACELP) Coding, March 1996
     Annex A

E.2 CS-ACELP Transfer Protocol

When the VoFR service user offers a frame of sampled speech it is immediately transmitted using the transfer structure described below.

E.3 CS-ACELP Transfer Structure

CS-ACELP produces 80 bits for each 10 ms frame of sampled speech. The list of the trans mitted parameters used by the CS-ACELP algorithm is provided below. In order to allow the frame relay device to adjust its transmission rate, the CS-ACELP transfer syntax structure will permit multiples of 10 ms frames to be packed into the voice information field. An integer number of 10 ms frames will be packed into the voice information field to form a $M*10$ ms payload. For each $M*10$ ms of compressed speech, $M*80$ bits or $M*10$ octets will be produced. Support of $M=2$ is required. A range of 1 to 6 can optionally be supported.

Symbol	Description	Bits
LSP0	Switched predictor index of LSP quantizer	1
LSP1	First stage vector of LSP quantizer	7
LSP2	Second stage lower vector of LSP quantizer	5
LSP3	Second stage lower vector of LSP quantizer	5
P1	Pitch period (Delay)	8
P0	Parity check of pitch period	1
C1	Fixed Code-Book – 1st sub-frame	13
S1	Signs of pulses –1st sub-frame	4
GA1	Gain Code-Book (stage 1) – 1st sub-frame	3
GB1	Gain Code-Book (stage 2) – 1st sub-frame	4
P2	Pitch Period (Delay) - 2nd sub-frame	5
C2	Fixed Code-Book – 2nd sub-frame	13
S2	Signs of pulses –2nd sub-frame	4
GA2	Gain Code-Book (stage 1) – 2nd sub-frame	3
GB2	Gain Code-Book (stage 2) – 2nd sub-frame	4
Total	Per 10 ms frame	80

*LSP = Line Spectrum Pairs

Figure E – 1 List of Transmitted Parameters

Octet	MSB	Bit Packing	LSB
1	LSP0, LSP1[7...1]		
2	LSP2[5..1], LSP3[5..3]		
3	LSP3[2,1], P1[7..3]		
4	P1[2,1], P0, C1[13..9]		
5	C1[8..1]		
6	S1[4..1], GA1[3...1], GB1[4]		
7	GB1[3...1], P2[5..1]		
8	C2[13...6]		
9	C2[5...1], S2[4..2]		
10	S2[1], GA2[3..1], GB2[4..1]		

Figure E - 2 CS-ACELP Bit Packing Structure for Each Frame

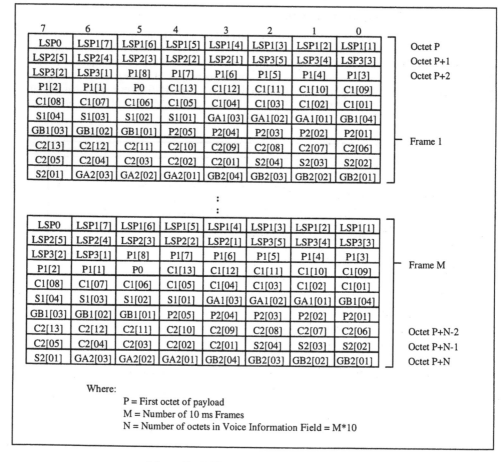

Where:

P = First octet of payload

M = Number of 10 ms Frames

N = Number of octets in Voice Information Field = M*10

Figure E - 3 CS-ACELP Transfer Structure

E.4   Transfer Characteristics

Packetization Time: M*10 ms

Algorithm Name	Reference Document	Compression Rate	Frame Size
CS-ACELP	ITU G.729	8 kbit/s	M*10

Figure E - 4 CS-ACELP Transfer Characteristics

E.5   Optional Sequence Number

Transmission of sequence numbers may be configured on a sub-channel basis. When enabled, the voice transfer syntax defined in Figure E - 3 is encapsulated in the Voice Transfer Structure field of the Active Voice Payload shown in Figure F - 2. The Sequence Number of Figure F – 2 shall be incremented every 10 msec. The Coding Type field of Figure F - 2 shall be set to 0000.

Annex F – Generic PCM/ADPCM Voice Transfer Syntax

F.1   Reference Documents

[1]   ITU G.711   Pulse Code Modulation of Voice Frequencies, 1988
[2]   ITU G.726   40, 32, 24, 16 kbit/s Adaptive Differential Pulse Code Modulation (ADPCM), March 1996
[3]   ITU G.727   5-, 4-, 3-, and 2-bits Sample Embedded Adaptive Differential Pulse Code Modulation, November 1994
[4]   ITU G.764   Voice packetization - Packetized voice protocols, December 1990

F.2   Voice Transfer Structure

Encoded voice samples – G.711 (PCM), G.726 (ADPCM), or G.727 (EADPCM) – sh all be inserted into the structure defined by Figure F - 1. The transfer of PCM/ADPCM/EADPCM is inspired by ITU-T Recommendation G.764. The following sections define two payload types and the Voice Transfer Structure.

Bit number	8	7		1	
MSB block	MSB/S8	MSB/S7	. . .	MSB/S1	P+ 5
			. .		
	MSB/S40	MSB/S39	. . .	MSB/S33	
MSB-1 block	(MSB-1)/S8	(MSB-1)/S7	. . .	(MSB-1)/S1	
			. .		
	(MSB-1)/S40	(MSB-1)/S39	. . .	(MSB-1)/S33	
. . .			. . .		
LSB block	LSB/S8	LSB/S7	. . .	LSB/S1	
			. .		
	LSB/S40	LSB/S39	. . .	LSB/S33	Octet N

Figure F - 1 PCM/ADPCM/EADPCM Voice Transfer Structure
(showing case of M = 1)

The voice transfer structure contains blocks arranged according to the significance of the bits. The first block contains the MSBs of all the encoded samples; the second contains the second MSBs and so on. Within a block, the bits are ordered according to their sample number. Since the 5 ms encoding interval corresponds to 40 samples, each block contains 5 octets.

A particular feature of this structure is that non-critical (enhancement) information is placed in locations where it can easily be discarded, without impacting the critical (core) information. For example, if 32 kbit/s EADPCM (G.727 (4,2)) is used, then there will be four blocks corresponding to four bits of varying significance (msb, msb-1, msb-2, lsb). The least significant blocks (msb-2, lsb) are the enhancement blocks and may be discarded under congestion conditions.

Annex G describes a related way of placing the critical and non-critical information into separate frames, so that the enhancement blocks can be marked with Discard Eligibility.

The size of the voice transfer structure depends on the packing factor M and the coding type, as shown in Figure F - 4. The packing factor is a multiple from one to 12. The value of M is configured identically at transmitter and receiver. It is typically, but not necessarily, the same in both directions. Equipment complying with this transfer syntax shall be configurable to support the default value M = 1.

When M is greater than 1, the voice transfer structure contains a first set of blocks, ordered from MSB to LSB, followed by a second set of blocks, ordered likewise, and so on up to the Mth set of blocks.

F.3   Active Voice Payload

When the Payload Type is Primary Payload, other fields in the sub-frame are as shown. The voice transfer structure containing encoded voice samples is defined in section F.2.

Bits

8	7	6	5	4	3	2	1	Octet
Sequence Number				Coding Type				P
Voice Transfer Structure								P+1

Figure F - 2 PCM/ADPCM/EADPCM Steady State Payload

F.3.1   Coding Type

The coding type field indicates the method of encoding PCM/ADPCM/EADPCM voice samples into the voice transfer structure.

The transmitting end-system shall only encode using algorithms for which there is decode support at the receiving end-system. The algorithms supported by the receiver are known by mutual configuration.

Values of the Coding Type field are defined in Figure F - 4.

F.3.2   Sequence number

The sequence number is used to maintain temporal integrity of voice played out by the receiving end-system. For PCM/ADPCM/EADPCM, the underlying encoding interval is 5 ms. Voice samples are processed with this periodicity and the sequence number is incremented by 1. After a count of 15 is reached the sequence number rolls back to 0.

The sequence number is incremented every 5 milliseconds, even when there is no active voice to be sent. This would be the case during a silence insertion period, if voice activity detection were operational.

The peer end-system expects to receive voice samples in sequence and within a certain time period. If voice activity detection is operational and no active voice is received, the peer end-system will continue to increment its expected sequence number every 5 ms.

When multiple voice samples are received in a single subframe (M > 1), the next expected sequence number is incremented by M.

F.4   Silence Insertion Descriptor (SID) Payload

When the Payload Type is Primary Payload with Silence Insertion, other fields in the sub-frame are as shown in Figure F - 3.

Bits

8	7	6	5	4	3	2	1	Octet
Sequence Number				Reserved				P
Reserved		Noise Level						P+1

Figure F - 3 PCM/ADPCM/EADPCM Silence Insertion Descriptor (SID) Payload

F.4.1   Reserved

This field is set to 000000 by the transmitter and is ignored at the receiver.

F.4.2   Sequence number

This field is the same as defined in F.3.2.

F.4.3   Noise Level

The background noise level is expressed in -dBm0. The receiver can use this field to play out an appropriate level of background noise in the absence of active voice.

Additional sub-frames of this type may be sent if the noise level changes or may be sent redundantly to increase the probability of being received.

This payload type should not be sent if voice activity detection is not operational.

F.5 Transfer Characteristics

Encoding interval: 5 ms

Packing factor: M = 1 to 12

Support of M=4 is required. A range of 1 to 12 can optionally be supported.

Coding Type	Algorithm Name	Reference Document	Compression Bit Rate (kbit/s)	Voice Transfer Structure (Octets)
0000	PCM A-law	ITU G.711	64	40*M
0001	"	"	56	35*M
0010	"	"	48	30*M
0011	PCM u-law	"	64	40*M
0100	"	"	56	35*M
0101	"	"	48	30*M
0110	ADPCM	ITU G.726	40	25*M
0111	"	"	32	20*M
1000	"	"	24	15*M
1001	"	"	16	10*M
1010	EADPCM (5,2)	ITU G.727	40	25*M
1011	(4,2)	"	32	20*M
1100	(3,2)	"	24	15*M
1101	(2,2)	"	16	10*M

Figure F - 4 PCM/ADPCM/EADPCM Transfer Characteristics

Annex G – G.727 DISCARD-ELIGIBLE EADPCM VOICE Transfer Syntax

G.1   Reference Documents

[1]   ITU G.727   5-, 4-, 3-, and 2-bits Sample Embedded Adaptive Differential Pulse Code Modulation, November 1994

G.2   Voice Transfer Structure

The voice transfer structure is the same as defined in Annex F.

G.3   Active Voice Payload

The G.727 EADPCM compression algorithm outputs core and enhancement information. This information is separately assembled into blocks. Core information is inserted into frames with low discard eligibility (DE=0), and enhancement information inserted into frames with high discard eligibility (DE=1).

Core and enhancement information, if required by a particular traffic type, may be combined within a single frame with DE = 0.

When the Payload Type is Primary Payload, other fields in the sub-frame are as shown in Figure G - 1. The voice transfer structure containing encoded voice samples is defined in Annex F.

Figure G - 1 shows only two sub-frames, one each for core and enhancement information, but transmitters are explicitly allowed to use the VoFR header to pack multiple sub-frames of the same kind of information into each frame, with DE = 0 or 1, correspondingly.

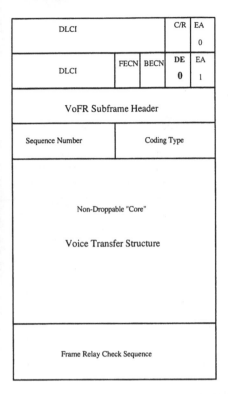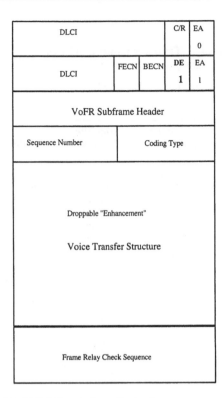

Figure G - 1 Discard-Eligible EADPCM Steady State Payload
(showing single sub-frames)

G.3.1   Coding Type

The coding type field indicates the method of encoding EADPCM voice samples into the voice transfer structure.

The transmitting end system shall only encode using algorithms for which there is decode support at the receiving end system. The algorithms supported by the receiver are known by mutual configuration.

Values of the Coding Type field are defined in Figure G - 2.

G.3.2   Sequence number

This field is the same as defined in Annex F.

G.4   Silence Insertion Descriptor (SID) Payload

This payload is the same as defined in Annex F.

G.5   Transfer Characteristics
Encoding interval: 5 ms

Packing factor: M = 1 to 12

Coding Type	Algorithm Name	Type of Information	Compression Bit Rate (kbit/s)	Voice Transfer Structure (Octets)
0000	EADPCM (2,2)	Core	16	10*M
0001	(3,2)	Enhancement	8	5*M
0010	(4,2)	"	16	10*M
0011	(5,2)	"	24	15*M
0100	(3,2)	Combined	24	15*M
0101	(4,2)	"	32	20*M
0110	(5,2)	"	40	25*M
0111	EADPCM (3,3)	Core	24	15*M
1000	(4,3)	Enhancement	8	5*M
1001	(5,3)	"	16	10*M
1010	(4,3)	Combined	32	20*M
1011	(5,3)	"	40	25*M
1100	EADPCM (4,4)	Core	32	20*M
1101	(5,4)	Enhancement	8	5*M
1110	(5,4)	Combined	40	25*M

Figure G - 2 Discard-Eligible EADPCM Transfer Characteristics

ANNEX H – G.728 LD-CELP Transfer syntax

H.1 Reference Documents

[1]   ITU G.728   Coding of Speech At 16 kbit/s Using Low-Delay Code Excited Linear
                   Prediction, November 1994

H.2 Voice Transfer Structure

Voice samples that are compressed using 16 kbit/s LD-CELP (G.728) will be inserted into the voice transfer structure defined in Figure H - 1. The LD-CELP compression algorithm produces a 10-bit code-word vector for every 5 samples of input speech from an 8000 sample/sec. stream. The 10 bits are reformatted to fit within the octet structure of the Voice Transfer Structure. Every group of five octets contains four, 10-bit code-words resulting in a 2.5 ms duration sub-frame. Two of these 2.5 ms groups are combined into a 5 ms block for transmission. The MSB of the first 10-bit code-word is aligned with the MSB of the first octet in the block. Subsequent bits of the code-word are placed in descending bit locations of the first octet with the other bits of subsequent code-words being bit packed into the remaining octets. Each block consists of eight 10-bit code-words which are mapped into 10 octets.

H.3   Transfer Protocol

9		3	2		0
MSB	7-bit shape vector	LSB	MSB	3-bit gain vector	LSB

(MSB)	7-bits of shape vector[0]	(LSB) (MSB) 1-bit of gain[0]	P
2-bits of gain[0] (LSB) (MSB)	6-bits of shape vector[1]		P+1
1-bit of shape vector[1] (LSB) (MSB) 3-bits of gain[1] (LSB) (MSB) 4-bits of shape vector[2]			P+2
3-bits of shape vector[2] (LSB) (MSB) 3-bits of gain[2] (LSB) (MSB) 2-bits of shape vector[3]			P+3
5-bits of shape vector[3]	(LSB) (MSB) 3-bits of gain[3] (LSB)		P+4
(MSB)	7-bits of shape vector[4]	(LSB) (MSB) 1-bit of gain[4]	P+5
2-bits of gain[4] (LSB) (MSB)	6-bits of shape vector[5]		P+6
1-bit of shape vector[5] (LSB) (MSB) 3-bits of gain[5] (LSB) (MSB) 4-bits of shape vector[6]			P+7
3-bits of shape vector[6] (LSB) (MSB) 3-bits of gain[6] (LSB) (MSB) 2-bits of shape vector[7]			P+8
5-bits of shape vector[7]	(LSB) (MSB) 3-bits of gain[7] (LSB)		P+9

Where: P = VoFR subframe header and optional transfer protocol octets

Figure H - 1 LD-CELP Voice Transfer Structure
(showing case of M = 1)

The size of the voice transfer structure depends on the packing factor M. The packing factor is a multiple from 1 to 12. The value of M is configured identically at transmitter and receiver. It is typically, but not necessarily, the same in both directions. Equipment complying with this transfer syntax shall be configurable to support the value M = 1 to 12.

When M is greater than 1, the voice transfer structure contains multiple blocks, starting with the first encoded voice sample and ending with the last encoded voice sample.

H.4   Transfer Characteristics

Encoding interval: 5 ms

Packing factor: M = 1 to 12

Other Capabilities:

In-Band Tone Handling - Can pass 2400 baud Modem Signals & DTMF

Algorithm Name	Reference Document	Compression Bit Rate	Voice Transfer Structure
LD-CELP	ITU G.728	16 kbit/s	10*M octets

Figure H - 2 LD-CELP Transfer Characteristics

H.5   Optional Sequence Number

Transmission of sequence numbers may be configured on a sub-channel basis. When enabled, the voice transfer syntax defined in Figure H - 1 is encapsulated in the Voice Transfer Structure field of the Active Voice Payload shown in Figure F - 2. The Sequence Number of Figure F – 2 shall be incremented every 10 msec. The Coding Type field of Figure F - 2 shall be set to 0000.

Annex I – G.723.1 MP-MLQ Dual Rate Speech Coder

I.1   Reference Document

[1]   ITU G.723.1   Dual Rate Speech Coder for Multimedia Communications Transmitting at 5.3 & 6.3 kbit/s, March 1996

I.2   Transfer Structure

Voice samples that are compressed using the 6.3kbit/s MP-MLQ algorithm (G.723.1 high rate) and 5.3kbit/s ACELP algorithm (G.723.1 low rate) yield a frame of packed parameters for every 240 samples of input speech from a 8000 sample/sec stream. Some of these parameters are based on an analysis of the entire frame; others are based on the analyses of each of the four component 60 sample sub-frames. Figure I - 1 shows of list of transmitted parameters for both MP-MLQ and ACELP.

For MP-MLQ, the resulting 191-bit frame is formatted to fit within the 24 octet structure of the Voice Information Field (one bit is unused) as defined in Figure I - 2. For ACELP, the resulting 160-bit frame is formatted to fit within the 20 octet structure of the Voice Information Field as defined in Figure I - 1. In Figure I - 2 and Figure I - 3, each bit of transmitted parameters is named PAR (x)_By: where PAR is the name of the parameter and x indicates the G.721 sub-frame index if relevant and y stands for the bit position starting from 0 (lsb) to the msb.

The expression PARx_ByPARx_Bz stands for the range of transmitted bits from bit y to bit z. The unused bit is named UB (value=0). RATEFLAG_B0 tells whether the high rate (0) or the low rate (1) is used for the current frame. VADFLAG_B0 tells whether the current frame is active speech (0) or non-speech (1). The combination of RATEFLAG and VADFLAG both being set to 1 is reserved for future use. Octets are transmitted in the order in which they are listed in Figure I - 2 and Figure I - 3. Within each octet shown, the bits are ordered with the most significant bit on the left.

Name	Transmitted parameters	high rate	low rate # bits
LPC	LSP VQ index	24	24
ACL0	Adaptive Code-Book Lag	7	7
ACL1	Differential Adaptive Code-Book Lag	2	2
ACL2	Adaptive Code-Book Lag	7	7
ACL3	Differential Adaptive Code-Book Lag	2	2
GAIN0	Combination of adaptive and fixed gains	12	12
GAIN1	Combination of adaptive and fixed gains	12	12
GAIN2	Combination of adaptive and fixed gains	12	12
GAIN3	Combination of adaptive and fixed gains	12	12
POS0	Pulse positions index	20*	12
POS1	Pulse positions index	18*	12
POS2	Pulse positions index	20*	12
POS3	Pulse positions index	18*	12
PSIG0	Pulse sign index	6	4
PSIG1	Pulse sign index	5	4
PSIG2	Pulse sign index	6	4
PSIG3	Pulse sign index	5	4
GRID0	Grid index	1	1
GRID1	Grid index	1	1
GRID2	Grid index	1	1
GRID3	Grid index	1	1

Figure I - 1 List of Transmitted Parameters

*Note: The 4 msb of these code-words are combined to form a 13 bit index, msb Position

TRANSMITTED	PARx By. ....
1	LPC B5...LPC B0, VADFLAG B0, RATEFLAG B0
2	LPC B13...LPC B6
3	LPC B21...LPC B14
4	ACL0 B5...ACL0 B0, LPC B23, LPC B22
5	ACL2 B4...ACL2 B0, ACL1 B1, ACL1 B0, ACL0 B6
6	GAIN0 B3...GAIN0 B0, ACL3 B1, ACL3 B0, ACL2 B6, ACL2 B5
7	GAIN0 B11...GAIN0 B4
8	GAIN1 B7...GAIN1 B0
9	GAIN2 B3...GAIN2 B0, GAIN1 B11...GAIN1 B8
10	GAIN2 B11...GAIN2 B4
11	GAIN3 B7...GAIN3 B0
12	GRID3 B0, GRID2 B0, GRID1 B0, GRID0 B0, GAIN3 B11...GAIN3 B8
13	MSBPOS B6...MSBPOS B0, UB
14	POS0 B1, POS0 B0, MSBPOS B12...MSBPOS B7
15	POS0 B9...POS0 B2
16	POS1 B2, POS1 B0, POS0 B15...POS0 B10
17	POS1 B10...POS1 B3
18	POS2 B3...POS2 B0, POS1 B13...POS1 B11
19	POS2 B11...POS2 B4
20	POS3 B3...POS3 B0, POS2 B15...POS2 B12
21	POS3 B11...POS3 B4
22	PSIG0 B5...PSIG0 B0, POS3 B13, POS3 B12
23	PSIG2 B2..PSIG2 B0, PSIG1 B4...PSIG1 B0
24	PSIG3 B4...PSIG3 B0, PSIG2 B5...PSIG2 B3

Figure I - 2 Octet Packing for the 6.3 kbps MP-MLQ codec

TRANSMITTED OCTETS	PARx_By, ….
1	LPC_B5…LPC_B0, VADFLAG_B0, RATEFLAG_B0
2	LPC_B13…LPC_B6
3	LPC_B21…LPC_B14
4	ACL0_B5…ACL0_B0, LPC_B23, LPC_B22
5	ACL2_B4…ACL2_B0, ACL1_B1, ACL1_B0, ACL0_B6
6	GAIN0_B3…GAIN0_B0, ACL3_B1, ACL3_B0, ACL2_B6, ACL2_B5
7	GAIN0_B11…GAIN0_B4
8	GAIN1_B7…GAIN1_B0
9	GAIN2_B3…GAIN2_B0, GAIN1_B11…GAIN1_B8
10	GAIN2_B11…GAIN2_B4
11	GAIN3_B7…GAIN3_B0
12	GRID3_B0, GRID2_B0, GRID1_B0, GRID0_B0, GAIN3_B11…GAIN3_B8
13	POS0_B7…POS0_B0
14	POS1_B3…POS1_B0, POS0_B11…POS0_B8
15	POS1_B11…POS1_B4
16	POS2_B7…POS2_B0
17	POS3_B3…POS3_B0, POS2_B11…POS2_B8
18	POS3_B11…POS3_B4
19	PSIG1_B3…PSIG1_B0, PSIG0_B3…PSIG0_B0
20	PSIG3_B3…PSIG3_B0, PSIG2_B3…PSIG2_B0

Figure I - 3 Octet Packing for the 5.3 kbps ACELP codec

I.3   Transfer Protocol

When the VoFR service user offers a frame of sampled speech it is immediately transmitted using the transfer structure described above in Section I.2.

I.4   Transfer Characteristics

Packetization Time: 30 ms

Other Capabilities:

In-Band Tone Handling - Can pass DTMF

Algorithm Name	Reference Document	Compression Rate	Frame Size
MP-MLQ	ITU G.723.1	6.3 kbit/s	24 octets
ACELP	ITU G.723.1	5.3 kbit/s	20 octets

Figure I - 4 MP-MLQ and ACELP Transfer Characteristics

I.5   Optional Sequence Number

Transmission of sequence numbers may be configured on a sub-channel basis. When enabled, the voice transfer syntax defined in Section I.2 is encapsulated in the Voice Transfer Structure field of the Active

Voice Payload shown in Figure F - 2. The Sequence Number of Figure F – 2 shall be incremented every 10 msec. The Coding Type field of Figure F - 2 shall be set to 0000.

Index